The Frontier Between the Map and the Territory

The Frontier Between the Map and the Territory

Notes on the Intertwining Between Matter and Information

By Manuel Varela

Figures by Jorge Sánchez

IFF BOOKS

London, UK

Washington, DC, USA

CollectiveInk

First published by iff Books, 2026
iff Books is an imprint of Collective Ink Ltd.,
Unit 11, Shepperton House, 89 Shepperton Road, London, N1 3DF
office@collectiveinkbooks.com
www.collectiveinkbooks.com
www.iff-books.com

For distributor details and how to order please visit the 'Ordering' section on our website.

ISBN: 978 1 80341 979 4 (Paperback)
978 1 80341 991 6 (ebook)
Library of Congress Control Number: 2024922824

A CIP catalogue record for this book is available from the British Library.

Design: Lapiz Digital Services

UK: Printed and bound by CPI Group (UK) Ltd, Croydon, CR0 4YY
Printed in North America by CPI GPS partners

To Coli Mestre, the real author behind the scribe

Contents

Acknowledgments

This book is the result of many hours of conversation with patients, colleagues, friends, and relatives. Acknowledging all of them would be impossible.

Borja Vargas, Jose Manuel and Jose Luis Agud, Laura Fernandez Sedano, Emilio García Delgado and Irene del Caso have read parts and offered helpful critics and comments.

My relatives have courageously withstood hours of obsessive discussions on these topics. Specifically, Ivan has read and commented on several preliminary drafts.

Robin Milner reviewed and polished up the English version.

Most of the ideas displayed in the book arise from many years of conversation with my wife, Coli Mestre, a clinical pathologist for over three decades. Pathologists are the definitive "baptizers" in Medicine, and thus are often painfully aware of the flaws and limitations of medical categories. Going through all these issues with her was like finding out that the Pope is agnostic: both unsettling and reassuring.

I. Introduction

The Map is not the Territory.

Alfred Korzybski

Why is it Urgent to Form an Opinion?

This text discusses the idea that reality has two polar opposite forms, which could in broad terms be called "map" and "territory." In the following sections, the characteristics of these forms will be analyzed in more detail, but it is important to emphasize the urgency of the discussion.

Perhaps never before has humanity been so immersed and intimately connected with technology while being so distant from the conceptual foundations that science handles (and that enable that technology). Our "common knowledge" about what time, space, or consciousness are is in radical dissonance with the concepts that science deals with (which enable the development of lasers, transistors, or atomic energy).

On the other hand, as frequently happens in science, technological developments precede and condition conceptual frameworks: the "boxes" within which we organize our world view. When this happens, ethical discussions often lag behind scientific development, responding reactively and with little capacity to lead the process of change. This is evident in areas such as eugenics or cloning, and it is probably inevitable.

However, we are approaching a threshold that could be particularly crucial, for which we have not even begun to prepare, and which might well overwhelm us in the coming years. This text does not intend to take sides or even present arguments for and against, but it is essential that we acquire the tools to form an opinion, because many of us will likely have to choose.

Transhumanism

In his book *The Singularity Is Near: When Humans Transcend Biology* (Kurzweil, 2010), Ray Kurzweil postulates that in the next few decades, technological advances will offer a form of immortality that will not be achieved through conventional medicine but through a process of "uploading" the individual (consciousness, memory, will...) into a computer. Kurzweil refers to this moment as the "Singularity," which he places around 2040.

Parallel to this, entrepreneur Elon Musk believes that Artificial Intelligence will become an unstoppable reality in a few years and that the "conventional" human species will quickly be surpassed from a cognitive perspective. Considering that cognitive capacity has been our main evolutionary tool, our future as a species will undergo unprecedented changes. The phenomenon may be similar to the appearance of *Homo sapiens* (and consider what happened to *Homo neanderthalensis*). Musk believes that our only option is to "jump on the bandwagon."

Our brain is, to some extent, an aggregation of superimposed structures:

- The rhombencephalon (medulla oblongata, cerebellum, and brainstem), the "reptilian brain," responsible for most automatic functions.
- A limbic system ("paleomammalian brain"), responsible for survival functions and basic emotions.

- A neocortex, responsible for most information processing, language, and consciousness.

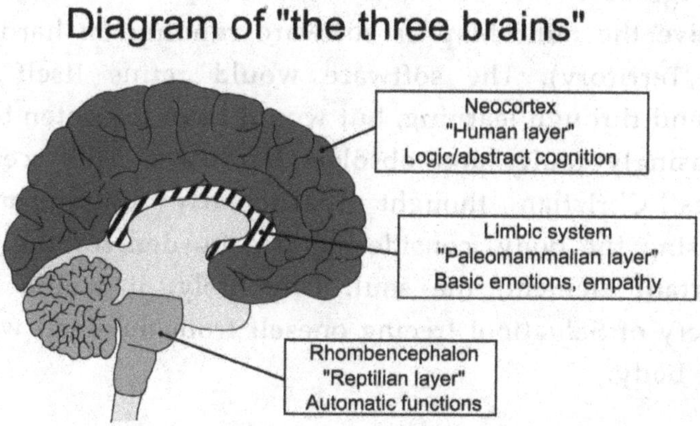

Fig I.0.1: Schematic structure of the brain.

Musk argues that we are starting to incorporate a new superimposed structure, upon which we increasingly depend for storing information, communicating, and forming ideas about our environment. This structure is digital (mobile devices, the Internet, prosthetics…). The only difference is that it is not "inside," and our bandwidth to communicate with this structure is ridiculously small (for example, typing speed on a mobile device). Musk believes that our only evolutionary option is to accept this new "neo-neocortex" and dramatically improve our bandwidth (Urban, 2017). In short, directly connecting the brain to the Internet.

In *Homo Deus* (Harari, 2016), Yuval N. Harari extensively discusses this scenario and emphasizes our "conceptual poverty" to handle it.

What unifies all these initiatives is nothing other than the migration from the Territory to the Map. As discussed in other sections, the interrelation between Map and Territory is not linear and often operates in leaps. However, this leap will be particularly critical and will mark a watershed moment in the history of humanity.

Religious echoes resonate in this initiative. It is easy to conceive the mind/Map as software running on hardware (body/Territory). The software would refine itself over time and through learning, but would have to contend with increasingly faulty and obsolete hardware. This scenario evokes Christian thought (particularly Calvinism) of despising the body, considering it a burden for the truly important element: the soul. Ultimately, it would be a mimicry of Salvation: freeing oneself from the dead weight of the body.

It is important to note that this is not science fiction. It is not even a hypothetical technological development. It is a real and growing movement. As of the time of writing these lines, Kurzweil is the Director of Engineering at Google and one of the executives (alongside Google and NASA) at Singularity University (https://su.org/), which aims to prepare for this scenario. Musk is a successful technology entrepreneur responsible for projects like PayPal, Tesla, and Space-X. He has also founded a company called Neuralink (https://www.neuralink.com/) precisely for this purpose. Facebook has changed its name (at present, Meta), and arguably its goals, with the intention of creating a Metaverse (virtual universe).

Perhaps the initial impulse of incredulity, of placing the topic in the realm of fantasy, is more of a defensive reaction than a rational judgment. As Max Tegmark (Tegmark, 2017) proposes, whether we like it or not, humans will have to decide whether to "transhumanize" or not. An interesting alternative is Reza Negarestani's neorationalism (Negarestani, 2018), discussed in section IV.2: "Transhumanism, Cis-humanism."

This Map/Territory dichotomy is ubiquitous, but one area where it is particularly evident is in clinical practice. All physicians have faced difficulties and misunderstandings that arise from the difference between the Map (the disease-diagnosis) and the Territory (the patient's ailments). Along with some incursions into the field of complexity analysis, this duality has confronted me for years. This text aims to provide some considerations for discussing the issue, addressing aspects from seemingly disparate fields. It is not possible to have exhaustive knowledge of all of them, and it may seem unwise (and perhaps border on charlatanism) to elaborate on these matters. However, all of these are nothing more than different aspects of the same dichotomy, and this multifaceted approach is essential to represent it.

Voices of criticism have been raised within the scientific community for some time (Bohm, 1980), pointing out that scientific endeavors increasingly focus on solving problems rather than providing explanations. This is particularly evident in Quantum Mechanics, which provides undeniable empirical results and forms the basis of much of our technology (e.g., semiconductors, transistors, lasers, information technology, imaging). However, Quantum Mechanics presents a reality that is counterintuitive and contrary to our "common knowledge," leading more and more scientists to abandon the pursuit of an "explanation of reality" and settle for objective and verifiable results that can solve specific problems. This is captured in the well-known quote attributed to Richard Feynman, Nobel laureate in Physics: *"If you think you understand quantum mechanics ... then you don't understand quantum mechanics."*

This relinquishment of explanation is accompanied by an increasingly esoteric language and conceptual instrumentation, resulting in the emergence of disconnected microworlds.

Occasionally, an idea jumps out of its microworld and successfully survives in other domains. However, this often comes at the cost of a profound deformation of the original concept, which greatly annoys the "priests" of the original microworld. This has happened with concepts such as chaos theory, quantum entanglement, and entropy.

Naturally, this problem is not new. The 19th century was marked by a controversy that pitted two opposing conceptions of University against each other (Berlin et al., 1999). The Enlightenment defended a university dedicated to society, believing that an important part of its function was to disseminate knowledge and advance the culture of civil society. In contrast, Romanticism saw the university as a "Temple of the Spirit" whose purpose was to produce science, with little concern for everyday life.

The current situation of our university is somewhat of a diluted hybrid. On one hand, it is seen as a factory producing technicians to feed the system (with little consideration for "knowledge irradiation" to civil society), and on the other hand, it has become an industry of curriculum production that is increasingly segmented and highly specialized.

Developing a multidisciplinary approach to the Map/Territory dichotomy is not easy, and it is fertile ground for speculations ranging from metaphors to charlatanism. This text attempts to address the topic from the perspective of what could be called "conventional science," although it inevitably touches upon fields which are not strictly scientific. Some sections may seem trivial to specialists in their respective disciplines, while others may not be of interest to many readers. However, I have tried to ensure that each section is readable on its own, allowing the reader to jump around and follow their own preferences.

The somewhat kaleidoscopic structure of this approach aims to demonstrate that various issues emerging from different areas of knowledge are manifestations of the same notion: the gap between two worlds that are immiscible yet inseparable.

The first part will discuss the general principles of the Map/Territory polarity. The second part will present some "border post" points where this dichotomy seems most evident. The third part will examine how this polarity affects biomedical fields and conventional clinical practice. Finally, we will explore how this problem has been central to Eastern philosophy for centuries and discuss some possible dilemmas and perspectives for the immediate future.

I.1: Map/Territory

This text aims to explore the idea that there are two polar opposite forms of being, which could metaphorically be called Map and Territory. Certainly, this is not new: it is an idea closely tied to dualistic conceptions that have traversed the entire history of Philosophy. However, it does not intend to engage in a philosophical discussion of the problem. This essay attempts to explore the Map/Territory pair in the light of the current state of science, as it might be perceived by anyone with basic scientific training, and at the same time interested in how technological advances (and especially Artificial Intelligence) are going to change our world.

One of the first formulations of this Map/Territory pair is offered by Plato in his *Republic*, when he separates the world of Ideas from the world of the Senses. Perhaps the most successful formulation is that of Descartes, in his classic dichotomy of *res cogitans* and *res extensa*. Naturally, this philosophical thought raised overwhelming problems, among others, the psychophysical problem (how a *res cogitans* without physical presence could interact with material reality) and has practically become the "enemy to beat" of current Neuroscience (Damasio, 2006).

But the Cartesian formulation was so successful that it has rooted in the collective unconscious to the point of becoming, due to its obviousness, virtually invisible. Thus, in Medicine we continue to separate organic disorders from mental ones, or we continue to avoid the contradiction presented by deterministic scientific thought with the problem of free will.

Problems of Definition

It is not easy to define precisely the terms of the Map/Territory pair, precisely because trying to define it means automatically placing oneself in "Map mode." However, in the absence of formal definitions, "archetypal examples" of this binomial can be explored.

Matter/Thought

This is the approach directly derived from the Cartesian perspective. The Territory would correspond to matter (*res extensa*), the Map would be thought (*res cogitans*). Naturally, all the classic problems remain. But an advantage of this formulation is that it avoids the dangers that have historically threatened dualistic thought (soul/body; God/Demon; Good/Evil), which has turned this conviction into a genuine "risky profession" that often led to the stake (Manicheans, Albigensians, Cathars...).

Matter/Information

One of the corollaries of the Cartesian approach is the need for a conscious observer. It would seem as if either an immanent soul is accepted, or a certain degree of neurological evolution is required for a *res cogitans* to appear. However, it could be argued that the Map (with some of the characteristics discussed later) could exist before an observer appears. An example would be the genetic code (and in general, replicating systems (Dawkins, 2016)), which appears as information that can be encoded in different substrates, transferable and reproducible independently of the material object in which it is embodied. In this sense, the Map would not depend on the observer, and would rather be the message: the encoded information (without consideration of the carrier element). In short, it would mean accepting that an algorithm exists in itself, independently of the matter in which it is embodied. We will explore the consequences of this approach (what Harari calls "Dataism" (Harari, 2015)) in other sections of this text.

Analog/Digital

The Territory would be characteristically analog. Thus, when we say that something measures 2 cm, we accept that with a sufficiently powerful instrument, this measurement would be corrected (2.000001 or 1.9999999). On the other hand, the Map is typically digital, and when we say 2 it is exactly 2.

Map objects are typically discrete (delimited, subject to categories and criteria), but when they are projected onto the Territory, "edge problems" systematically arise: Territory objects are not discrete, and the limits (and therefore the interface with the Map object) become complicated. It is almost impossible to find a "Territory object" with sharp edges. The digital world can get as close as it wants to the analog world (paradoxically, the digitization of sound has greatly increased the accuracy of reproductions), but there will always be a gap (as small as desired) separating the analog Territory from the digital Map.

Wave/Particle

Quantum physics posits a dual reality, in which any particle can be in an "analog" wave form, diffusely distributed following Schrödinger's equation, to some extent "Territory," until observation collapses the wave function, "digitizing" the particle in a certain position: in some way conceptualizing it and turning it into "Map."

There are countless manifestations of the Territory/Map binomial (a dish and its recipe, hardware versus software, music and its score, etc.). In each case, the leap from Territory to Map entails a selective extraction of the essential characteristics, which are then encoded in a way that facilitates storage, transmission, and assembly. Naturally, this implies a loss of the rest of the characteristics. The consequences of this loss will be a recurring theme in this text.

Different Worlds, Different Laws

In some way, the Map and the Territory obey different laws.

- The very logic is different: the Map world follows an "Aristotelian" logic, of excluded middle ("A is B or

A is not B: there are no other alternatives") with well-defined categories. The Territory world does not follow this principle, and judgment is conditioned by an ever-changing context: an 8-year-old individual can be considered a man (versus an animal) or a non-man (child versus adult); a brain-dead patient is a man (biological species), or he is not (can be used as an organ donor). This logical discrepancy has been a central theme in disciplines such as Cognitive Psychology (Rosch, 1978) or General Semantics (Weinberg, 1991) and is behind many of the taxonomic problems in the Natural Sciences (see section III.6: "Sex, Race, Species, Gene").

- The Territory obeys causes, while the Map is guided by reasons. The Map is deterministic (e.g., in a two-dimensional system, the square of the hypotenuse of a triangle is exactly the sum of the squares of the other two sides: there are no nuances). In the Territory, it is not possible to offer a closed causal argument (processes are multicausal, there is the phenomenon of sensitivity to initial conditions, it is impossible to completely isolate the context, etc.).
- The model of reasoning is different in the Map and in the Territory (see I.2: "Cognitive Processing in the Map and the Territory"). In the Territory, heuristic reasoning prevails, allowing for incomplete induction ("Whenever I have seen A, B occurs. It is predictable (but not certain) that this A will also be followed by B."). The causal reasoning in the Map is algorithmic and is based on evidence and non-contradiction (Postulate: if A -> B, when A -> not B is observed, there is something wrong).
- The Map follows the principle of Recursive Decomposition: a problem can be recursively divided into many subproblems. Once each of the subproblems is worked out, the system can be reassembled, and the initial problem is solved. In the Territory, this is not always the case. The whole can be

more than the sum of its parts (or it becomes impossible to make a "clean dissection" of the subproblems). In any case, there are properties of the whole that the analysis of the parts does not explain (emergent phenomena: it is impossible to predict the chemical characteristics of water based on the characteristics of oxygen and hydrogen; it is impossible to predict the behavior of an animal based on its chemical composition).

- The Territory world is limited to three spatial dimensions (the issue of time is dealt within section II.1: "Time"). The Map world does not have these limitations and can operate perfectly in an n-dimensional space.
- Many physical variables decay in the Territory world following the inverse square law (light, sound, gravity…).[1] This law of decay does not operate in the Map world. As a first approach, the transmission of information could be considered dichotomous: either it arrives or it is corrupted. It is possible to "analogize" the degree of corruption, in order to assess the extent to which the information contained dissipates. In section II.8: "Renormalization and Complexity," we will see that this is one of the interfaces between map and territory and is closely related to concepts such as entropy or renormalization techniques.
- The Territory world exists in space: it is typically a "here and now." In contrast, the Map world lives fundamentally in time. It inevitably has a spatial dimension (which generally concerns its "territory" aspect, where it is embodied: cartography, recipe, equation, plan, etc.), but its raison d'être is to endure over time.

Abstraction as a Leap from Territory to Map

Abstraction and categorization processes are examples of the leap from Territory to Map. Through these processes, certain characteristics of the Territory are selected as relevant, while others are omitted. At the cost of some "impoverishment" and decontextualization, a "standardized" Map object is created, which can be packaged and stored (concept), transmitted (word), or assembled (argumentation). The process is somewhat similar to digitizing a figure: an image is "simplified" through pixelation, resulting in a digital file.

Naturally, this offers enormous possibilities that are not available to "Territory objects." The cost of this leap will be one of the recurring themes of this text.

General Semantics (Weinberg, 1991) identifies several distinct levels in this leap. For example, think of an apple:

- The first level is the (presumed) physical object-phenomenon, inaccessible, only hypothesized (at the end of the process).
- The next level would be our organism's physical reaction to incoming stimuli from the apple (tactile, visual...). This triggers physicochemical reactions (release of neurotransmitters, configurational changes in the neuronal network...), and a series of action potentials that travel to the central nervous system (thalamus, primary sensory cortices), where these sensations become part of consciousness.
- Finally, there is a step of constructive elaboration in relational areas, that produces the concept of "apple," ready to connect with other concepts or be transmitted.
- Even though the (presumed) objective physicochemical reality is analog, it is important to understand that our first contact with it is already digitalized. The entire neurosensory and conceptual elaboration system is based on neural networks (Fig. I.1.1) (in the strict sense of interconnected nerve cells), and each neuron functions as an analog-digital decoder.

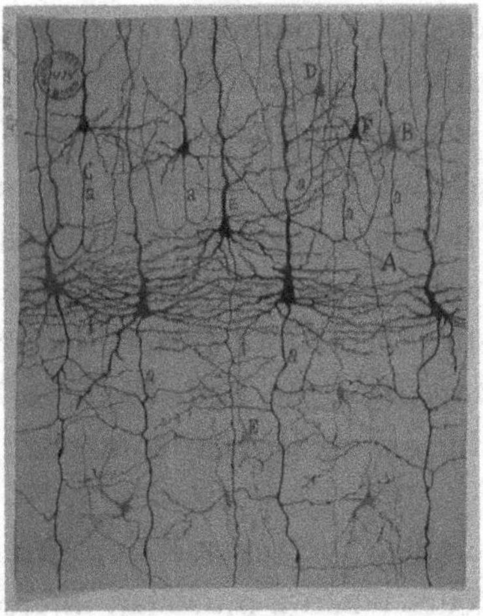

Fig. I.1.1: Drawing of the cerebral cortex structure, by Ramón y Cajal. https://www.unesco.org/en/memory-world/archives-santiago-ramon-y-cajal-and-spanish-neurohistological-school

Sensory neurons detect stimuli (which can be analog) and, based on a calculation of stimulating or inhibitory afferences (i.e. a nerve impulse arriving to a neuron), discharge (or do not discharge) an action potential. These action potentials impact upon other neurons, which compute the activating or inhibitory afferences and in turn "decide" whether or not to trigger an action potential (which, depending on the neurons, can be stimulatory or inhibitory). This means that not only the final process of abstraction is a form of digitization: our contact with the (presumed) objective reality is digitized from its very inception. It is noteworthy that the specific function of the cell that forms what is probably the most complex system known is simply to act as an analog/digital decoder, that is, to be a bridge (or a boundary) between the Territory and the Map. We have no pure analog experience: from the first moment our experience is digitized. Interestingly, mystical thought posits that intellectual knowledge will always be mediated, and that direct knowledge requires different mechanisms to access reality.

It should be noted that this model is based on the conventional conception of abstraction as the cognitive process by which we create a Map by selecting certain characteristics extracted from the Territory. In section I.3: "Mind/Map versus Brain/Territory" we will see that this is a huge oversimplification, and that, in many respects, the actual process is the reverse: we only perceive those characteristics of the (supposed) Reality that are able to penetrate our Map-filter. As Seth (Seth, 2021) suggests, our perception is nothing more than a "controlled hallucination."

Naturally, there is also the reverse step, from the Map to the Territory: this occurs when a musician interprets a score, or a dish is cooked from a recipe, or a house is built from an architect's plan, or a nuclear reactor is constructed based on the theory of Relativity. Edward A. Lee suggests that the fundamental difference between Science and Engineering is that Science tries to make a model that explains how reality works (leap from Territory to Map), while Engineering uses a model to construct an instrument that solves a specific problem (leap from Map to Territory) (Lee, 2017).

Advantages of the Map

In exchange for the simplification/reduction it implies, the Map mode has some undeniable advantages:

The Map as a "Layer Dissection" of the Territory

The Map is in some way an aggregate of different "layers" of the Territory. Therefore, one of its great advantages is that it allows for individualizing and manipulating specific aspects of the Territory in isolation, without interfering with the rest. Let's consider the pair "score/musical interpretation" as an example of Map/Territory. The score has several "layers" upon which one can act specifically:

- Melody (sequence of notes)
- Rhythm (relative duration and beats: whole, half, quarter notes, 2/4, 6/8...)
- Harmony (key, chords...)
- Timbral color (flute, violin...)
- Tempo (*Adagio, Allegro, ritardando*...)
- Intensity (*piano, forte, crescendo*...)
- Intention (phrasing, *legato, staccato*...)

It is relatively simple to modify each of these aspects in the score without altering the others. This would be much more difficult when operating "as a whole" on the Territory.

Computational Advantages

Grouping multiple aspects (sensory, mnemonic, etc.) into a single object (concept) greatly facilitates computational processing. If a computational conception of the mind's functioning is accepted, the emergence of conceptualization processes seems evolutionarily obvious.

Combinatory Advantages

Discrete and well-defined elements can be combined (e.g., phonemes for words, or words for ideas), exponentially increasing their power.

Robustness of Digitization

Another advantage of the transition from analog to digital is the increase in stability. The "contextual noise" required to change a "0" to a "1" is much greater than that needed to modify an analog value. Consequently, the robustness of the digital signal (in its persistence over time or in its transmission) is much greater than that of an analog signal.

Indexing

One of the main advantages of Map information is that it is indexable, and therefore very easy to search or reorder for different purposes. From a computational standpoint, this is an unbeatable advantage.

Overcoming Spatial-Temporal Constraints

The Map is a selective extraction of the information present in the Territory. This allows it to be freed from its original context and transported or transmitted in any other digital matrix. Likewise, it allows it to be stored and preserved "out of time."

Generalization

The selective extraction of information allows for generalizations and the application of this information in contexts similar, but not identical, to the original.

The Territory/Map Polarity as an Archetype

The Map/Territory dichotomy is ubiquitous. However, there are some areas where it is especially striking, and some disciplines have even been established precisely on this dichotomy. Other sections analyze some of these cases (e.g., II.2: "Entropy(ies)"; II.3: "Differential Calculus"; II.8: "Renormalization and Complexity"; I.4: "Mind/Map versus Brain/Territory"; III.3: "Statistics," etc.). But in addition to the Aristotelian distinction between matter/form, or the Cartesian *res extensa/res cogitans* duality, echoes of the Territory/Map polarity can be seen in multiple systems of very diverse origins:

Eastern Philosophy

IV.1: "Resonances from the East" comments on the similarities between this Map/Territory dichotomy and some lines of Eastern thought. In very general terms, Taoism postulates the existence of two principles (Yin and Yang) (Wang, 2012). The Yin principle would have a contextualizing, grounding, synthetic component: what we could metaphorically consider Territory. In contrast, Yang would be essentially focusing, analytical, proactive: what metaphorically corresponds to Map.

Hinduism postulates Brahman as the universal principle and ultimate Reality (Territory) and Atman as the characterizing or individualizing principle (Map) (Lochtefeld, 2002).

Popper: Worlds 1, 2, and 3

Karl Popper postulates the existence of what he calls three worlds.

- World 1 would be physical entities: the Territory in the strict sense.
- World 3 would be *"the world of the products of the human mind, such as stories, explanatory myths, tools, scientific theories (whether true or false), scientific problems, social institutions, and works of art"* (Popper & Eccles, 1977). It's a good description of what we call the "Map" in this text.
- Finally, Popper defines a world 2, which he describes as *"the world of mental states, including states of consciousness, psychological dispositions, and unconscious states"* (Popper & Eccles, 1977). As discussed in I.4: "Mind/Map versus Brain/Territory," this would be one of the clearest examples of the boundary between Map and Territory, modelling the interface between those two forms of reality.

Bohm: Explicated Order/Implicated Order

To try to explain some characteristics of conventional quantum mechanics, which in his view were inadmissible (especially having to admit an irreducible, spontaneous randomness), the theoretical physicist David Bohm proposed a model of reality that could be described as an infinite sea of energy (implicate order), with some islets where this implicated order "coagulates" into an existence in time and space (explicate order) (Bohm, 1980). This dualism of implicate/explicate order bears clear parallels with the Territory/Map polarity.

Psychoanalysis: Id/Superego

Psychoanalysis postulates a structure of the psyche that could be described as being composed of:

- an impulsive, irrational Territory-instance, guided by the impulse of immediate pleasure, without temporal nuance (the id)
- a structuring, rational Map-instance, subject to norms that connect with its space and time (the superego)
- a "boundary" that negotiates between these two instances (the ego); I.3: "Mind/Map versus Brain/Territory" (primary and secondary consciousness) explores some neurophysiological hypotheses concerning this model.

Genesis

Genesis (Genesis, 1989a) speaks of an initial Golden Age in the Garden of Eden, without consciousness of time (or at least without the idea of death), with the prohibition on taking the fruit from the Tree of Knowledge of Good and Evil. However, the serpent tells Eve *"...when you eat from it your eyes will be opened, and you will be like God, knowing good and evil."* Eve understands that it was *"desirable for gaining wisdom,"* so she takes the fruit and offers it to Adam. Thus, they realize their nakedness and become aware of their mortality. In some way, this can be considered as a leap from the Territory (existence in a continuous present) to the Map (measurable time, abstract judgments).[2]

Two Worlds, Two Hemispheres

The ubiquity of this polarity brings to mind the collective unconscious posited by Jung and forces us to consider to what extent this polarity is related to our neurological structure. Although it is an oversimplification, it is accepted that the left hemisphere is predominantly verbal/conceptual (Map), and the right hemisphere is predominantly contextualizing (Territory). Naturally, this does not clarify the causal direction:

- It is possible that this seemingly ubiquitous polarity is merely a consequence of our two ways of processing information (left hemisphere versus right hemisphere).
- Or it might be the other way around: our divided neurological structure is a successful evolutionary response to a real polarity. Certainly, it seems that responding to this polarity (the basis of the cognitive revolution) has been one of the most successful evolutionary mechanisms of the human being.

Footnotes

1. An interesting exception is frequency, which does not decay with distance. This may be explained by this being a "Map-variable" from the outset. Frequency is expressed in Hertz (cycles/second), and cycles are measured in whole numbers; thus frequency is "born discretized."
2. One cannot avoid thinking about transhumanism when, later on, Jehovah says, *"The man has now become like one of us, knowing good and evil. He must not be allowed to reach out his hand and take also from the tree of life and eat, and live forever ... After he drove the man out, he placed on the east side of the Garden of Eden cherubim and a flaming sword flashing back and forth to guard the way to the tree of life."*

I.2: Cognitive Processing in the Map and the Territory

Cognitive processing also differs between the Map and the Territory. These can in broad terms be called analytical processing and holistic processing.

Analytic Processing ("Map")

The analytic model focuses on objects, which it considers as independent entities organized into categories with their own "essence." It assumes that there is a deep structure to reality that obeys certain stable rules and laws that can be investigated and learned. The first of these laws would be Logic (in essence, the principle of non-contradiction). A sufficiently precise analysis of reality would allow for a reductionist approach: problems can be divided into sub-elements, which, once resolved one by one, can be reconstructed to solve the overall problem. These tools foster the development of Theoretical Science, which allows for making predictions. If at any point empirical findings do not match these predictions, it is necessarily concluded that either the data are erroneous or the rule being used was imperfect and must be corrected.

The object/logic approach of analytic processing also extends to systems and ideas: it considers a network as a set of nodes interconnected by lines. Learning/teaching is fundamentally algorithmic, based on categories and laws, making it relatively "mechanizable" and very easy to disseminate.

Nisbett et al. (Nisbett, 2001) propose classical Greece as a model (and perhaps the origin) of this cognitive schema. This was a society organized into relatively small, not isolated but independent units, in which the individual (citizen) had special preeminence. Individual freedom was assumed, with the capacity to choose (and consequently, responsibility for what was chosen). This society gave birth to many of the revolutionary ideas that have shaped our current Western society, and which, because they are so evident, have become invisible (e.g., the idea of "essence"; the idea of "Nature"; or the idea of "freedom").

Holistic Processing ("Territory")

The holistic model focuses on relationships rather than objects. It does not posit an "inherent essence," and organizes reality more by associations and similarities than by taxonomies. It emphasizes impermanence and does not fix categories and laws definitively. Its main norm is more about harmony than logic, and its approach to unforeseen events is more dialectical. When faced with contradictions, it maintains a suspended judgment and seeks a "third way."

Holistic processing does not assume that a problem can be dissected into perfectly independent parts: it would consider a network as a set of lines/relationships whose intersections constitute the nodes. Learning/teaching is fundamentally heuristic, based on tangible data and experience, making it much less systematizable.

This model has been very prevalent in Eastern thought, and Nisbett (Nisbett et al. 2001) cites Imperial China as an example of a society with a vast and complex structure in which the relevance of the individual was conditioned by the harmony of the whole. Perhaps this cognitive model explains the modest development of metaphysics and Theoretical Science in the Eastern world. Also, skepticism towards a reductionist attitude may explain why Eastern medicine shunned surgical approaches.

Table I.2.1

Analytic processing	Holistic processing
Focused on objects	Focused on relations
There is a fixed, stable reality	Awareness of impermanence: only change is persistent
Search for rules and laws	Search for similarities and relationships
Reductionistic approach: problems may be decomposed and solved in parts	Gestalt approach: elements of a problem are not independent
Logic as main criterion	Harmony as main criterion
Algorithmic learning	Heuristic learning
Classical Greece as cultural model	Imperial China as cultural model
Net as a set of connected nodes	Net as a set of lines (nodes are just crossing points)

Teaching and Managing Information

These differences are clearly reflected in the ways of teaching and learning. Learning in the holistic model is essentially heuristic, based on experience and tangible data through similarity or opposition. It's how children learn: *"this is a cat; and this and this...; this, however, is a dog"* (Rosch, 1978). This makes it slow and laborious to acquire. However, once learned, the access and use of that information is immediate and effortless ("intuitive knowledge"). In contrast, the analytical model is very systematizable and therefore easy to explain and learn. However, access to information is slower and requires effort.[1] We naturally find it much easier to work intuitively in a "holistic mode," allowing for quick and effortless choices (Kahneman, 2012). Instead, the "analytical mode" requires a "gear shift" and involves will and effort. Probably in many fields (and certainly in Medicine), one of the hallmarks of excellence is knowing when to "change gears."

Example of Heuristic versus Algorithmic Learning: Pulse Exploration

Pulse exploration is an essential tool in Eastern medicine (Kuriyama, 1999). Its learning takes years (to the point that it is not even attempted in introductory courses to Eastern medicine), and its description is characteristically metaphorical (floating pulse, hollow, as a drum skin, slippery, etc.). Conversely, Western medicine takes an essentially algorithmic approach: it just counts the beats per minute. Naturally, this is very objective, minimally dependent on the observer, and extremely easy to teach. As expected, its use quickly became widespread. It is true that some specialists include nuances (arrhythmic, slow, bounding, etc.), but these characteristics are becoming increasingly obsolete, and the essence of taking the pulse (beats/min) is already taught in secondary education centers.

Heuristic Learning: Vocal Education

An example of heuristic learning is the acquisition of language. Similarly, in vocal music education, instructors must resort to an entirely metaphorical language (e.g., "speak as if with a hot potato in your mouth," etc.) to teach voice placement and use of resonators. Probably this area is much less "algorithmizable" than other physical activities because the anatomical-physiological categorization is much less developed than in the rest of the body. While there is an exhaustive description of the musculature of the neck, torso, and limbs, the musculature responsible for phonation is practically undescribed. This is arguably because, even during the height of the Anatomoclinical school (see section III.2: "Medical Narratives"), autopsies systematically omitted dissection of the face for instinctive reasons of respect for individuality. In any case, the structural and functional knowledge we have about the musculature of phonation is rudimentary.

Footnote

1. Consider the following example (extracted from Kahneman, 2012): A baseball set, consisting of a bat and a ball, costs 11€. We know that the bat costs 10€ more than the ball. How much does the ball cost? The immediate, heuristic (and incorrect) response is almost always 1€. (The correct answer is, obviously, 0.5€.)

I.3: The Great Journey

The history of humankind is, to a certain extent, the story of a journey from the Territory to the Map, or at least an unstoppable development of the Map within the species *H. sapiens*. Likely, this tendency to explore the Map-world is driven by physiological mechanisms. The human brain is an excellent pattern recognizer (the first step towards making an abstraction), and one of the most effective gratifications is precisely the identification of a pattern. This gratification is behind such essentially human aspects as innate curiosity, scientific interest, guessing games, or puzzles. The ability to recognize patterns is one of the basic tools of creativity, ingenuity, or humor, and certainly increases the fitness (and probably the reproductive capacity) of the possessor. In reality, identifying a pattern is nothing more than drawing a map of a territory area. In essence, it's about building a bridge between the two worlds.

The Cognitive Revolution: Language as a Bridge between the Territory and the Map.

A crucial step in this journey from the Territory to the Map occurs around 150,000–200,000 BC (Pagel, 2017). For reasons which are unclear (but probably include genetic mutations in the FOXP2 gene (Fisher & Marcus, 2006)), a language emerges that is tremendously more flexible than that of other species, which conditions the capacity not only to sound an alarm or to pass specific information (which exists in other species) but also to formulate abstract concepts. This represents a clear tool to leap from the Territory to the Map: it allows for the selective extraction of features, "standardizing" them and connecting them for later use.

It could be said that this is not an exclusively human property, and that other animals also have language and communication capacity. However, human language has a characteristic typical of the process of jumping between the Territory and the Map, which it does not share with any other animal. Human phonation involves a simplification of the laryngeal anatomy compared to primates (Nishimura et al., 2022). This implies a greater "standardization" of phonemes, which allows their connection and the use of a combinatorial system, greatly expanding its power. This combinatorial ability has not been found in other animals.

This strategy of *{simplification -> standardization -> combinatorial method}* is ubiquitous in the jumps from the Territory to the Map, and is often observed in Nature as a strategy to manage diversity. Some obvious examples are the genetic code, based on the combination of 4 bases (adenine, guanine, cytosine, and thymine), or the production of the entire range of proteins from 20 amino acids, or the synthesis of an almost unlimited number of antibodies from a small set of genes (Helbert, 2017).

Language, as a step towards the Map, allows the development of complex codes and the organization of relatively large social structures, allowing its members to collaborate towards a common goal (hunting, group defense), and gives the group an unbeatable evolutionary advantage. Consequently, this revolution is accompanied by an unprecedentedly rapid expansion of *H. sapiens*, and the rapid extinction of all other Homo species.

From the Agrarian Revolution to the Industrial Revolution

The transition from a hunter-gatherer society to an agrarian society is accompanied by a significant increase in the size and complexity of social organizations. It should be noted that the replacement of hunter-gatherer societies with agrarian societies is not necessarily "good" in the sense of being favorable to its members. There is evidence suggesting that this transformation was accompanied by a decrease in life expectancy and worse nutrition (Bowles, 2011; Diamond, 2017). This evolution is due to strict "Darwinian" reasons: agrarian societies were more fertile, and the gradual increase in population was displacing previous social structures. It is important to bear in mind that the Darwinian mechanism does not apply only to genes. Dawkins (Dawkins, 2016) defines the idea of a replicator as a structure that is capable of making sufficiently faithful copies of itself to maintain stability, but allowing some "copy errors" that lead to minimal variability, and consequently, to competition among copies. When these conditions are met in a world of limited resources, the structure automatically enters the "Darwinian ratchet" of natural selection. This naturally applies to genes, but also to ideas (memes) or cultures. And the competition between replicators (fitness) is measured exclusively in their success of propagation. As we will see, this also applies to models of society. Applied to religions, it explains the almost constant persecution by religions of birth control systems or non-reproductive sex.

Writing

The development of increasingly larger communities demands the creation of new "Map-objects" for intersubjective use: religions, rituals, hierarchies, currencies… With the advent of writing, these innovations can crystallize and become independent of memory and immediacy (and to a certain extent, of humans) (Goody, 1977). Writing is another example of the model *{simplification -> standardization -> combinatorial method}*. Attempting to directly represent ideas leads to hieroglyphics, immensely complex with high computational and memory costs. However, by standardizing phonemes and using combinatorics, it is possible to exponentially increase power while maintaining a very limited set of symbols. In any case, writing comes with a profound modification of the concept of time. Hunter-gatherer societies lived in a continuous present, virtually without accumulation processes. In contrast, the agrarian society is forced to foresee the future (adjusting to seasons for sowing or harvesting, accumulating for winter or times of scarcity, paying taxes). Territory-time (kairos) begins to transform into Map-time (chronos) (see II.1: "Time").

Influence of Christianity on (Self-Styled) "Developed" Societies

An important step in this journey from the Territory to the Map is likely conditioned by Christianity (Henrich, 2020). The "Darwinian evolution" of religions has tended to select monotheistic creeds, with a supreme, male, and omnipotent god. Among them, the Christian religion was particularly successful, perhaps because it incorporated three aspects that now seem evident to us but were revolutionary at the time: (1) it offered a life beyond death, depending on behavior in this life; (2) it considered man as a subject capable of free choice (and therefore ultimately responsible for his actions); and (3) it established a universal morality that affected all men, regardless of their creed ("natural law," the Ten Commandments).

Additionally, it established a series of rules that had enormous social consequences, probably unpredictable and not obvious at first glance. Among them, in addition to prohibiting polygamy, it forbids marriage between cousins. For many centuries, this prohibition extended to sixth cousins. This fight against intra-clan marriages is even reflected in the language (the spouse of a sibling is called in English "brother-in-law," referring to the incest taboo). In a predominantly rural society with very small populations, this forced people to seek spouses often far from their birthplace. The Church benefited as well because it was possible to obtain an exemption from this rule (in exchange for money, land, or other goods). Naturally, this exemption was practiced almost exclusively among royalty and nobility.

The prohibition of marriage with close relatives results in the destruction of a social structure based on clans. Historically, clans (essentially familial) were the cornerstone of society. They were groups of people who lived in a relatively small common territory, with somewhat fluid borders, but which encompassed a relatively extensive family structure. Being forced to seek a spouse outside the clan initiated a series of social changes that ultimately conditioned a qualitatively different psychology and socio-political structure. Among the induced changes are:

- Newlyweds tend to move to another domicile.
- Individual property is consolidated (as opposed to collective property by the clan). An individual inheritance system is established.
- Intra-clan bonds weaken rapidly, and the focus falls more and more on the individual. Loyalty to the clan is replaced by loyalty to general principles (justice, individual responsibility, reliability). Shame (which extends to the clan) is replaced by guilt (which affects the individual). The analysis of infractions is increasingly based on individual psychology (intentionality, character, and history of the offender) and less on the objective consequences and responsibility of the clan.
- A "chronos time" common to all individuals and independent of vital time is consolidated. Punctuality appears as an important virtue, and one can "lose time," "gain time," or think that "time is money."

This evolution is also seen in the shift from a holistic, contextualizing cognitive processing, very much based on relationships with the environment, to analytical thinking, principlist, based on abstract concepts (freedom, rights, justice, equity). Some "group" virtues (group loyalty, respect for authority, obedience) are replaced by others more centered on the individual (independence, creativity, honesty).

New forms of organization of individuals appear, who voluntarily group around their own ideas or objectives (monastic orders, universities, guilds). This is fertile ground for the development of commerce and thereby the interconnection between unrelated individuals, which must rely on an external code accepted by both parties. Finally, the Protestant Reformation actively promotes literacy (it considers it the obligation of every faithful to read the Bible personally) and collaborates in the development of compulsory education. This path of individual consolidation continues with the Enlightenment, the French Revolution, Human Rights, or the Industrial Revolution. Ideas of equality and freedom lead to a democratic political model (theoretically), which spreads from Europe and North America to the rest of the world.

This is not simply a plausible story. Joseph Henrich (Henrich, 2020) provides ample empirical data demonstrating the close statistical correlation between the historical influence of Christianity in different geographical areas with multiple variables demonstrative of this evolution (incidence of marriage between cousins, multiple tests exploring various aspects of individualistic focus, political systems, analytical versus holistic cognitive processing, etc.).

This development of the Map-world over the centuries has been based on the production of more and more intersubjective objects ("Map-objects"), which only exist to the extent that their meaning is shared (currency, nation, market, charters, contracts), and which allow the organization of ever larger structures. Initially, they were cities, then empires, and finally, we are reaching a globalized world, which shares increasingly similar "Map-objects." This ability to "externalize" and share Maps is an extraordinary "evolutionary accelerator." The fact that several minds can co-work on the same instruments (concepts, words, ideas, formal education) exponentially increases their power and evolutionary speed. More recently, technological development has made it possible to carry out these tasks online, overcoming spatiotemporal constraints and further accelerating the transformation.

As with all processes of transition from the Territory to the Map, this leap implies a certain loss of detail or characteristic features, in exchange for the possibility of connecting these objects together and growing the mega-structure.

Dataism

Perhaps the next step on the path to the Map is Transhumanism (and its theoretical correlate: Dataism). This current of thought understands the "self" (self-awareness, emotions, memories, values, will...) as an algorithm implemented in a system (the central nervous system), and postulates that it is precisely the architecture (the "form/software," and not the "matter/hardware") that endows it with its special characteristics. Consequently, if a sufficiently precise copy could be made, it would be possible to "duplicate" the algorithm in another system and therefore "transplant" consciousnesses. Moreover, the system would not even have to be a brain: it could create artificial consciousness.

To oppose this view, one would have to postulate the existence of some specific internal or external element that infuses the algorithm with its "conscious" characteristics (somehow, the equivalent of a "soul"). The empirical data supporting this option are scarce or nonexistent.

In its most radical version, Dataism posits itself as the next evolutionary step. Just as chemistry encompasses but goes beyond the physics of elementary particles, and biology encompasses but goes beyond chemistry, Dataism would be the next step that encompasses and goes beyond biology. This conception, which at first glance could be considered "anti-humanist," nevertheless has significant precursors. If instead of labelling it as an "algorithm," it is called "spirit," Dataism is the direct heir of the German Idealism of the 19th century (Schelling, Hegel) (Marías, 1971a; Negarestani, 2018).

There are several lines of argument against Dataism:

- The first would be based on technical difficulties. Currently, these are insurmountable: we would need to know in real-time the precise situation of each of the brain's connections, constantly in motion. And even if this were possible, there would be a phenomenon of "sensitivity to initial conditions" (Wikipedia, Chaos Theory) that would prevent an identical copy. Within this argumentative line would be the model of consciousness proposed by Antonio Damasio (Damasio, 1999), discussed in I.5: "Consciousness as a Map." This model suggests that the dichotomy between mind/software and brain/hardware is an oversimplification. Consciousness would require a neurohumoral framework (the proto-self) that is not easily reducible to an algorithm, and therefore not "digitizable." In any case, it's a weak defense: it could produce a "self" perhaps not identical, but with similar characteristics. On the other hand, this does not dismantle the possibility of a convergence between biological and synthetic brains, as proposed by Neuralink (https://www.neuralink.com/).
- A more conceptual attempt to differentiate between consciousness and an algorithm is Searle's Chinese Room thought experiment (Cole, 2015). (Very broadly, the argument proposes an individual locked in a room, with all the necessary books and unlimited time. He would receive a question, written in Chinese, and would have

> to search the books for an answer to that question, which he would transcribe (also in Chinese) and send outside. The argument points out that, although the algorithm works correctly, the subject of the experiment is not aware of the question or the answer.) This argument has sparked a long controversy. Probably no one questions the ability of an Artificial Intelligence to memorize, accumulate experiences, make judgments (informed by values), make decisions, and execute them. What is at issue is the possibility that this AI has consciousness and intentionality. Ultimately, the question is whether semantics is simply the consequence of correct and sufficiently complex syntax. This topic is discussed more extensively in IV.2: "Transhumanism, Cis-humanism."

Finally, there is a parallel (and perhaps more unsettling) line of argument. If the real "hard problem" is consciousness, nothing would prevent the construction of "intelligent" systems without consciousness. Probably, a 19th century German idealist would argue that consciousness (the "spirit") would be an emergent property of the algorithmic development itself, but a "radical Dataist" could suggest that consciousness is simply an evolutionary selected property in humans because it has facilitated their capacity for organization at a historical moment when data processing was very rudimentary. Once a sufficient data processing capacity and bandwidth are reached, the "Dataist systems" might not need this property.

The Evolution of Power

The evolution of the sources of social power can also be read in terms of the journey from the Territory to the Map.

Initially, power was strictly physical, nuanced by instrumental aspects (mastery of metal technology) and organizational (group, clans). Subsequently, the source of power was the domain over land, and this situation persisted until the end of the Middle Ages.

However, the progressive development of commerce and navigation gave rise to a bourgeoisie that controls the financial framework and becomes an increasingly decisive agent. The Industrial Revolution completes this transformation, and the origin of power shifts to the possession of the means of production.

In a more recent phase, power seems to be shifting towards the means of exchange. The economy becomes increasingly financial, and the value of a company depends not so much on its production or dividends as on its share price.

As we approach the present, the view is necessarily less objective and more subject to opinion. The conservative vision would argue that the development of democratic systems has led to an increasingly stable system, with the Market as an instrument capable of optimizing the production and distribution of goods. A more radical view would postulate that the holders of the means of production and exchange have taken control of the political system and the media. This hegemony is now fundamentally cultural and does not allow for the conception of another possible organizational regime, although its future is bleak (climate crisis, migratory, technological, etc.).

Probably, technological development is conditioning a new change in the sources of power. It is becoming increasingly clear that the new "gold" is information, and above all, the control of its distribution.

In any case, it seems difficult to deny a shift in the sources of power from an essentially Territory-world to an ever more Map-world.

Language at the Different Stages of "Life"

Language is a reflection of the leap from the Territory to the Map. Although language has been a crucial element in the development of these map-objects, nuances of "territory" can still be observed in natural language, which are lost in the formal language (mathematics, computing) inherent to digitalization. Both are ways of encapsulating information and allowing its transmission. But natural language (more "Territory") is characteristically context-dependent. George Lakoff (Lakoff & Johnson, 2013) posits that this is because natural language is based on metaphors related to how we interact with our environment. These metaphors are not just "tools to communicate with one another": they are primarily ways of understanding, and they constitute the foundations of our conceptual edifice. In natural language, there is no objective truth, and the criterion of truth (the comprehension and acceptance of the message) is coherence with the underlying (and, in principle, shared) metaphorical system. The statements "your theory has a solid foundation" (metaphor of theory as construction) or "I am wasting my time" (metaphor of time as a quantity) make sense, but "a time with solid foundation" or "losing a theory" would be nonsensical. This dependency on the metaphorical context is what allows for sarcasm or irony. Natural language always has a context (and sometimes a nonverbal metalanguage). In contrast, computer language cannot be ambiguous or polysemic. Each element can only have a single meaning, which remains the same regardless of context, and has a criterion of "objective truth": the absence of contradiction with the axioms that underpin it.

This difference between natural language and formal language (the former dependent on a metaphorical model, the latter on an axiomatic set) makes it difficult for us to imagine what the language of evolved Artificial Intelligences will be like. Tegmark (Tegmark, 2017) proposes a differentiation of life into 3 stages:

- Life 1.0: essentially basic biological life (animals, plants). They are not capable of modifying either their software or their hardware, which only change subject to the (very slow) Darwinian natural selection mechanism.

- Life 2.0: life with a cultural dimension. Although there is evidence of a certain cultural dimension in animals (Whiten, 2021), this dimension is essentially human. The hardware is still subject to Darwinian evolution, but they can modify their software. This produces a kind of Lamarckian mechanism that greatly accelerates their evolutionary capacity.
- Life 3.0: "technological" life: advanced Artificial Intelligences. They can modify both their software and their hardware. Presumably, this will entail another leap in evolutionary speed.

Language at the level of life 1.0 is a tool for coordinating motor actions (e.g., giving an alarm signal, indicating where there may be food). In life 2.0, language serves to coordinate the coordination of motor actions (Echeverría, 2003) (organizing a hunt, conveying intentionality, agreeing on symbolic content). It is not clear that we can imagine the function of language in life 3.0. As Lakoff (Lakoff & Johnson, 2013) suggests, it is impossible for us to conceptually manage our environment without resorting to metaphors related to how we interact with it. These metaphors are so embedded in our language that they seem inescapable, and this probably explains some of the paradoxes that science confronts us with (e.g., the need to change our concepts of time and space).

It is difficult to imagine how Artificial Intelligences that do not rely on these metaphors will "think." Furthermore, do we understand what it might mean to coordinate the coordination of the coordination of motor actions?

I.4: Mind/Map versus Brain/Territory

One of the crucial border crossings between the Map and the Territory is, of course, the brain. This is where sensory inputs are recorded and where the "maps" we use are generated. It is also the ground where the classic psychophysical problem was raised: how can a nonmaterial entity (the mind) act on matter?

The following sections discuss two aspects:

- Why does the brain have this seemingly unstoppable need to produce maps that explain reality?
- What could the neuroanatomical basis of this process be?

General Outline

A general outline of the journey would be:

1. The idea of existence as a struggle against entropy.
2. The Free Energy Principle. Minimization of predictive errors as a surrogate for entropy.
3. The brain as a Bayesian device.
4. Hierarchical Predictive Coding as the brain's mode of operation.
5. Psychedelic drugs as an example of manipulation between predictions and stimuli.
6. The Default Mode Network, the Freudian ego, and primary and secondary consciousness.
7. A revision of our concept of time in light of predictive processing.

What Do We Mean by "To Exist"?

If we drop a drop of ink into a glass of water, the drop disappears. If we drop a drop of oil, the drop remains: it continues to exist.

What defines existence? A possible approach would be: a stable separation between "what is" and "the outside," with a defined interface (for example, the bilipid layer of a membrane). Formally, this interface acts as a Markov blanket: what remains inside can be defined by itself without requiring information from the outside. The model also assumes the existence of processes in which the outside acts on the inside, and processes in which the inside acts on the outside.

Another condition for existence is a certain persistence in time. Or, put another way, avoiding a "phase transition," a change significant enough to consider that the system has ceased to exist (e.g., the dissolution of the drop of ink, the dissolution of a snowflake, or the transition from life to death). This is why one of the essential characteristics of living systems is homeostasis, which at its core is just an attempt to limit changes, to restrict the number of states visited by the system – in summary, to limit entropy (in the informational sense) and to maintain some structure.

This entropy-based approach allows the borrowing of a whole set of theoretical instruments from Thermodynamics and Information Theory, among them, the Free Energy Principle. This is a theoretical construct that demonstrates that the persistence of a dynamic system in time requires a continuous process of minimizing a variable called Free Energy (see below). In evolutionary terms, only systems capable of minimizing their Free Energy survive over time (they are "selected"). More precisely, persistence over time is essentially minimizing Free Energy.

Self-Information, "Surprisal"

Imagine a system that can be in a multitude of possible states (S_1, S_2, ... S_i), with a certain probability (p_{s1}, p_{s2}, ... p_{si}) for each of these states. The *self-information* or *surprisal* of each state is defined as the base-2 logarithm of the probability of that state (with inverted sign):

$$\text{Surprisal of } S_i = -\log_2(p_{si})$$

The higher the probability of S_i, the lower its *self-information* or *surprisal* (the more "predictable" S_i would be).

The system's *entropy* would be the average of the *surprisal* associated with each state. Therefore, minimizing *entropy* (that is, persisting over time) would be reducing the average *surprisal*.

The problem is that, naturally, the subject does not have access to information about these states.

Free Energy Principle. The Bayesian Brain.

This is where the concept of Free Energy comes into play (Friston, 2009, 2010; Friston et al., 2010; Mann et al., 2022). It is an Information Theory concept derived from Statistical Physics and was developed to tackle complex probabilistic problems. Essentially, it is an upper limit to *surprisal*. Consequently, minimizing Free Energy is equivalent to minimizing surprisal (and consequently, entropy).

The fundamental advantage is that, while information about surprisal is not accessible to the subject, information about Free Energy is. To calculate surprisal, the subject would need to know the "real" probabilities associated with each state, which he has no information about. However, Free Energy can be evaluated by knowing sensory states (which are accessible) and a *recognition density*. *Recognition density* is a probability distribution of estimated causes associated with sensory stimuli (e.g., the probability that an external condition "x" produces a sensation "a"). This can be encoded in the subject's internal states and is therefore accessible. This model allows us to conceive of the brain as a Bayesian device, composed of two essential (and updatable!) "databases":

- The aforementioned *recognition density* (the probability that an external condition "x" produces a sensation "a").
- Some *priors*, which would be a distribution of the estimated probabilities of possible causes (e.g., how likely is the external condition "x").

The perception process would consist of comparing sensory stimuli against these two data sets. At its core, it solves the problem:

1. Given the probability of existence of "x" (which I can access through my priors),
2. and the probability that "x" produces a sensation "a" (which I can access through my *recognition density*),
3. when I encounter a sensation "a," how likely is it that I am in the presence of "x"?

Following the classic Bayesian model, this would produce a *conditional density* or *posterior density* that explains the sensations received (Fig. I.4.1).

	PROBABILITY OF PRODUCING SUCH A SOUND	PROBABILITY OF MEETING IT	POSTERIOR DENSITY
HORSE	0·8	0·4	0·32
ZEBRA	0·8	0·0001	0·00008
DOG	0·1	0·6	0·06
HUMAN	0·05	0·8	0·04

Fig. I.4.1: A very simple example of Bayesian processing. When I hear clip-clopping in an urban context, I consider the probabilities of certain animals making that sound (Recognition Density), the probabilities of encountering such an animal (Priors), and from the intersection of both, I deduce the most likely origin of the sound (Posterior Density); in this case, a horse.

The hierarchical architecture of the brain would also allow it to evaluate the difference between predictions and the sensory stimuli received (*prediction errors*). It can be demonstrated that minimizing Free Energy equates to minimizing this prediction error and consequently correcting/adjusting the probabilities of the baseline model. In summary, **persisting over time means being able to minimize prediction errors.**

The system can reduce its prediction errors (and consequently minimize its Free Energy) through two mechanisms:

- Adjusting its model (refining its priors and/or recognition density). This would be achieved through a Hebbian learning system ("neurons that fire together wire together," consolidating circuits). Consequently, when there is a good correlation between predictions and inputs, the model is strengthened. This concept resonates with neural Darwinism, which posits that neural circuits are selected based on their "value" (= frequency of use). This is the underlying mechanism for perception (in terms of interpretation, as opposed to sensation), memory, and learning (the more times a prediction is confirmed, the more confidence I have in it).
- Selective sampling, "in favor of the model," confirming predictions (for example, the person who walks down a staircase blindly, trying to reach with their hand where they *know(?)* the handrail should be). In summary, seeking "self-fulfilling predictions." This is likely a potent natural selection mechanism, favoring sensory systems that align with the model and deselecting those that are not coherent. It's the classic "use it or lose it" principle of natural selection: useful systems hypertrophy, while useless ones atrophy.

Thus, the Free Energy model suggests that the brain (and generally all biological systems, and perhaps even all systems in the broadest sense) operate along a gradient that tends to reduce differences between their sensory stimuli and predictions. This must operate over a wide range of timescales (from milliseconds to years), using various mechanisms. Thus, neuronal activation mechanisms (in the millisecond range) would be used for rapid adjustments (sensations, movements), facilitation or inhibition mechanisms of connections or neurohormonal systems for systems operating in minutes to days (attention, perception), or synaptic or neuronal remodeling methods for longer processes (memory, learning). One could even postulate a similar model over very long timescales, subsuming the evolutionary mechanisms of natural selection: fundamentally, Darwinian adaptation could be interpreted as a gradient toward minimizing prediction errors.

This architecture might also explain the transformation of continuous to discrete variables that accompanies the leap from Territory to Map (see Analog/Digital, in I.1: "Map/Territory"). Cognition would be based on a recursive process, comparing *priors* with *posterior densities* at each step. As with all recursive loop models (such as the Turing machine), this would require step-by-step operation and therefore force a "discretization" of the variable. It also aligns well with the binary structure ($\log_2$) of usual entropy formulations: at each step, the model decides whether the adjustment is better or worse.

This model is also coherent with the Map/Territory polarity: the brain is a constant generator and optimizer of Maps. Its capacity to persist in the broadest sense — its resistance to disappearing (avoiding a "phase transition"), its adaptability, and its evolutionary fitness — depends on this capability.

An important nuance is that the evolutionary pressure to decrease Free Energy tends to minimize prediction errors, but this does not automatically imply an alignment with "Reality" (whatever that term means). It only implies a gradient toward reducing differences between recruited stimuli and predictions, attempting to constrain the model's complexity (more on this later).

Hierarchical Predictive Coding

The brain essentially functions as a producer of "maps," constantly comparing them to the "territory" to improve its predictions (sampling "in favor" or adjusting its *priors* and/or *recognition densities*). However, the functioning of these processes is not trivial and yields a rather counterintuitive picture of the perceptual process.

We tend to think of perception as a process of receiving stimuli that ascend from receptors to progressively "higher" areas of the cortex, being processed along the way. Thus, the eyes would be the "windows" through which the mind sees reality.

But this is not how perception works (Clark, 2016b). The brain is constantly producing predictions (based on previous inputs). The perceptual process is structured in several superimposed layers, organized hierarchically, where "ascending" stimuli are faced with "descending" predictions. What ascends to the next level is not the "corrected stimulus," but rather the prediction error: the difference between what is "expected" and what is "found." Conscious perception is achieved only when these prediction errors are "explained away," that is, when an adequate correlation is reached between descending predictions and ascending stimuli. This represents a Copernican shift in the perceptual mechanism, shattering our conventional conception of cognitive processing and implying that perception is fundamentally a controlled hallucination (Clark, 2016b). An example of this mechanism can be seen in the hollow-face illusion (https://michaelbach.de/ot/fcs-hollowFace/index.html). Although we are perfectly informed that the image is the reverse of a mask, our perception "knows" that faces are convex, making it impossible to see the "real" image of the concave face. Another example is the Ames window (https://www.facebook.com/watch/?v=255140663211496).

A similar model applies to motor activities. Performing a voluntary movement would consist of "imagining" (predicting) how the proprioceptive sensors would be in the desired position and adjusting motor plans to reach that "prediction."

One consequence of predictive processing is that it blurs the boundaries between sensory and motor aspects, incorporating them into a single recursive loop. This is especially evident in vision (Clark, 2016a). The idea of a retina that passively captures light impulses to transmit them to higher levels is entirely naive. The perception mechanism (sensory) is intimately linked to saccadic movements (motor), and drawing a line between sensory and motor would be entirely artificial.

There's also a parallel circuit to the upward↔downward traffic, where, at each level, a range of "reliability" or precision (inverse of variance) is attributed to sensory stimuli, determining to what extent *prediction errors* (differences between stimuli and predictions) deserve consideration (and should be raised to the next level) or can be dismissed as irrelevant. This "potentiometer," which regulates the relative weight of stimuli versus predictions, is closely related to the attention mechanism, which, when activated, tends to overvalue ascending afferents (making it harder to dismiss unexpected differences).

New Disease Mechanisms

It has been speculated that this alteration in the balance between ascending stimuli and descending predictions might be a common pathophysiological mechanism for several diseases. For example, it could offer a unifying explanation for seemingly different symptoms such as the hallucinatory and delusional features of schizophrenia (Adams et al., 2013). Thus, dysregulation of the ascending/descending flows at lower levels (close to the stimuli) could facilitate hallucinations (seeing or hearing complex stimuli that nobody else perceives), while dysregulation at higher (conceptual) levels could allow the development of delusional syndromes (aberrant ideations that the subject considers obvious, but no one else shares). Ultimately, in both cases, it's about an "error in map/prediction formulation" that is unable to "filter out noise" and sends to the upper level prediction errors that need to be explained. There is solid empirical data supporting this dysregulation between ascending and descending flows (Adams et al., 2012; Fletcher & Frith, 2009).

Similarly, this dysregulation (essentially, the inability to "filter out" prediction errors that physiologically would be considered "irrelevant noise" and therefore be cancelled) might explain some characteristics of autism (Pellicano & Burr, 2012), conversion disorders (Edwards et al., 2012), phantom limb syndrome (De Ridder et al., 2014), restless leg syndrome (Hanzlíková et al., 2019), the placebo effect (Büchel et al., 2014), or fibromyalgia.

Psychedelics

Hierarchical predictive coding allows the explanation of many of the phenomena associated with psychedelic drugs (de Caso, 2022). These drugs are specific agonists of a subtype of serotonin receptor (5-HT2AR), especially present in the upper associative cortex and notably in a network of structures called the Default Mode Network (DMN; see more below). Treatment with psychedelics characteristically depresses the "descending" flow of priors, weakening predictions, and consequently facilitating the ascent of unresolved prediction errors. This reduces "certainties" and leaves open many more possible interpretations of sensory stimuli, which would explain the hallucinatory symptoms associated with these drugs. More generally, the mechanism of these drugs could be explained as an increase in the brain's entropy, in the sense of increasing the possible brain states (Carhart-Harris & Friston, 2019). A metaphorical representation (Fig. I.4.2) could be a topography with valleys and mountains, where consciousness moves through the landscape, influenced by external stimuli but also subjected to the terrain and the force of gravity, tending to occupy the valleys (the deeper, the more likely). Psychedelics would act by "flattening" the landscape, thus facilitating movement and visits to other states, less accessible under normal conditions.

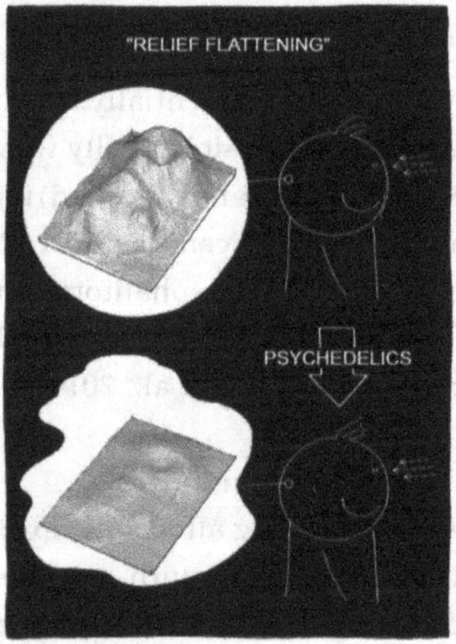

Fig. I.4.2: "Relief Flattening" as a representation of the entropy increase caused by psychedelic drugs

Another metaphorical image would be to consider the mind as a liquid surface on which sensory stimuli fall like rain. Psychedelics would act by reducing the liquid's viscosity, allowing the ripples triggered by each "sensory drop" to move and spread more easily.

Criticality

The brain is an example of a dynamic system operating in a type of regime known as the "edge of chaos" (https://en.wikipedia.org/wiki/Edge_of_chaos). These dynamics are characterized by being far from thermodynamic equilibrium, consuming significant amounts of energy, and being in a state that is close to, but not strictly, chaotic. This dynamic region has very special properties. Among others:

- It is exquisitely sensitive to perturbations.
- It can spontaneously organize into transiently stable states (metastable).
- It produces "avalanches": abrupt state changes that cascade throughout the system.

The entropic brain hypothesis (Carhart-Harris, 2018; Carhart-Harris et al., 2014) posits that the central nervous system operates in a subcritical dynamic regime (close to, but below, a truly chaotic regime). One way to evaluate the degree of (sub) criticality would be to measure its entropy. As mentioned in II.2: "Entropy(ies)," the term entropy generates significant conceptual controversies, and there are multiple metrics to evaluate it. However, there is robust evidence correlating a decrease in consciousness with a drop in entropy, and some of these techniques are routinely used to assess the depth of anesthesia (Roche & Mahon, 2021). Administration of psychedelics characteristically increases the system's entropy (Carhart-Harris & Friston, 2019), consistent with the "landscape flattening" image, increasing the number of "visitable states."

A critical component in this process appears to be the Default Mode Network (DMN).

Introducing the DMN

The brain appears to have two opposing neural systems (Fox et al., 2005): one focused outward (mostly activated when the mind is engaged in specific tasks or external stimuli) (Task-positive Network, TPN), and another mainly focused "inward," which is inhibited when the mind is focused "outward" and activated primarily at rest (Default Mode Network, DMN) (Raichle & Snyder, 2007). These two systems are clearly negatively correlated: the more active one is, the more inhibited the other is. The DMN has some striking characteristics:

- It receives more blood flow and has a much higher metabolic consumption than the average for the brain.
- It is composed of highly connected centers, yet relatively distant from primary sensory and motor centers.
- It is a system that is intensely activated when the subject engages in metacognitive tasks (self-reflection, theory of mind (e.g., "What do these people think about me?"), time travel (e.g., "What was I like as a child? What will I be like as an old man?"), contextualization, moral judgments).
- Its connectivity progresses from childhood to adulthood.
- There is evidence suggesting an expansion of the DMN system throughout evolution (Van Essen & Dierker, 2007).

These characteristics suggest the DMN could be in charge of general coordination tasks for brain function, aligning with what has been called secondary consciousness.

Primary and Secondary Consciousness

Although a clear dividing line cannot be drawn, it is accepted that there are two distinct forms of consciousness (Carhart-Harris et al., 2014):

- A primary consciousness that relates to sensory, perceptual, and mental imagery contents, but without time awareness.

- A secondary consciousness involving time management (memory/past, future/plans), abstraction, metacognition ("consciousness of being conscious"), moral judgments, and volition.

The DMN seems closely tied to the emergence of secondary consciousness, both in development (in each individual) and evolutionarily (in the species). Although there is no strict boundary between primary and secondary consciousness, it is assumed that they are qualitatively different and imply a kind of "state change" that seems (both evolutionarily and in the individual) linked to a sufficient development of the DMN. This network reinforces *priors*, limiting visitable states ("sharpening the landscape," Fig. I.4.2) and reducing entropy. This would presumably reduce "surprisal," increase prediction precision, and promote more focused, realistic, and foresighted thinking. In return, it would lose creativity (ability to think "outside the box") or flexibility to see new viewpoints. In a way, it's shifting from an "exploring brain" to an "exploiting brain."

Apart from early childhood (when secondary consciousness has not yet developed), the typical waking state is one of secondary consciousness. However, there are situations where regression to the primary state occurs. The clearest examples are the REM phase of sleep, states associated with temporal lobe epileptic seizures, and perhaps some extreme stress situations or the early stages of certain psychotic disorders.

Carhart-Harris and Friston (Carhart-Harris et al., 2014) suggest that a clear example of regression to primary consciousness is the effects of psychedelics.

A phenomenon frequently associated with regression to primary consciousness is the decoupling between the DMN and the medial temporal lobe nuclei. This phenomenon is observed in crises originating in temporal lobe foci but is also typical in psychedelic states (Carhart-Harris et al., 2014; de Caso, 2022).

The development of the DMN might result from pressure to minimize Free Energy, limiting entropy and reducing the system's degree of criticality. This is assumed to bring adaptive advantages by increasing foresight and realism. Conversely, primary thinking would be less subject to the downward pressure of predictions, allowing "unresolved prediction errors" to reach the consciousness, leading to alternative interpretations more influenced by desires or fears (i.e., magical thinking, wishful thinking). This aligns with the psychedelic effect ("psychedelic" comes from *psyche*, mind, and *diloo*, to open or manifest). The psychedelic state would have more open and creative thinking (but be less organized and planning-oriented).

This could have medical implications. Certain depressive, obsessive-compulsive, or addictive syndromes might be considered of "low entropy" (tendency to visit the same mental states), and treatment with psychedelics have yielded extremely good results (de Caso, 2022; Dyck, 2005; Grob et al., 2011; Krebs & Johansen, 2012; Moreno et al., 2006; Vargas et al., 2020).

This narrative of mental maturation from childhood as a control over primary impulses has clear Freudian resonances, and the DMN has been proposed as a structural correlate of the Freudian ego (Carhart-Harris & Friston, 2010).

The relationship between the "anti-entropic" effect of the DMN and its role in time consciousness (as an essential feature of secondary consciousness) is also intriguing. Again, we slide over the slippery concept of entropy, but it's worth remembering that entropy is what defines the arrow of time, marking irreversibility. If we ultimately have to abandon the idea of objective time external to the observer (see Hyperpriors in IV.2: "Transhumanism"), we might need to revisit the DMN with fresh eyes.

Conclusion

From all the above, the image of the brain emerges as a factory continually generating Maps, with a "management system" (establishing the relative importance of stimuli and predictions) in which the DMN seems heavily involved.

There would be selective pressure to optimize the alignment between Map and Territory, in the form of a tendency to minimize Free Energy. But an important nuance is that the minimization of Free Energy (i.e., survival) does not depend on its alignment with "Reality" (whatever that may mean...), but on minimizing prediction errors (the difference between what the senses (ascending) recruit and what is "expected" (descending)), trying to constrain the model's complexity. And this implies a balance between efficiency and thoroughness. Ensuring very high alignment (overfitting) involves greater computational expense and higher dependence on small details (with the consequent risk of lower capacity for generalization and application to similar but not identical situations). It seems that one of the functions of sleep may be "synaptic pruning" to optimize the predictive model (Tononi & Cirelli, 2006).

From an evolutionary point of view, it's not so much about "making the predictive map more realistic" as about improving fitness (optimizing reproductive success). And these need not always go hand in hand: a schematic but effective model may be preferable to a more perfect but more complex model (more on this in II.8: "Renormalization and Complexity"). For example, Donald Hoffman (Hoffman, 2016), based on both Neuroscience and Quantum Mechanics, suggests that certain aspects we consider "primary" or irreducible, like space or time, may be very useful simplifications, but it's important not to consider them as literally real. Hoffman proposes a metaphor of the icons on our computer desktop. The icon that represents a file is "real" and can be manipulated. But it would be an error to believe it is physically a small square we can pick up with the mouse and throw in the trash. It is a "shortcut" to simplify and avoid all the underlying computation. To Hoffman, space and time serve a similar role: they help organize and manipulate reality, optimizing our survival, but it would be a mistake to take their existence literally. Here we can also see the resonances of 18th–19th century German idealism. As with Kant's noumenon, these positions distance us from the possibility of capturing objective reality, and like Kant's *a priori* intuitions, they remove space and time from reality and place them in the observer's eyes. This approach also aligns with the Eastern idea of Maya (illusion, unreality) (https://en.wikipedia.org/wiki/Maya_(religion)).

The model also ties into the computational theory of the mind ("Strong Artificial Intelligence"), which postulates that consciousness is merely the result of an algorithm. After all, either we locate consciousness in some kind of objective entity (which would be, in a way, the heir to the soul, for which there's no evidence), or we attribute it to a certain form of organization. As Tegmark (Tegmark, 2017) says, consciousness is simply how information feels when processed in a sufficiently complex way.

Obviously, this would be another nail in the coffin of anthropocentrism: we're not at the center of the Universe, nor does the Sun revolve around us (Copernicus), nor are we the pinnacle of creation (Darwin). We would simply be an algorithm running on some computing system. Naturally, this need not necessarily be our brain. If our perception (everything we see or know, including that we have a brain, which seems the seat of knowledge) is the result of an algorithm, replicating it would be trivial. And by pure statistics, our chances of being a replica (not the original computer of the algorithm) would be astronomical. The last blow to anthropocentrism: we would end up being a virtual replica of an executing algorithm (although it would make no sense to ask where or when). Again, echoes of Genesis and the expulsion from Paradise...

This also illustrates the precise but indescribable fractal structure of the boundary between the Map and the Territory: we thought we were building the Map from the Territory (conceptualization, "discretization" paying a price in decontextualization), and we find ourselves back in the Platonic stage: the Territory is the embodiment of another Map, without concretizable space or time.

Of course, this doesn't explain or resolve (or contradict) the Map/Territory dichotomy, but it's an example of how these two ways of being are articulated. Although it seems like the Map aspect ("Information") is gaining prominence compared to a Territory ("Matter") that is increasingly elusive, this *pas-de-deux* evokes the relationship between Brahman and Atman in Hindu thought (see Duality: Conjunction of Opposites in IV.1: "Resonances from the East").

I.5: Consciousness as a Map

The Problem of Consciousness

One of the central problems in neuroscience is consciousness, which has been termed the "hard problem" of this field. A way to describe it is through a classic thought experiment. Let's imagine a future neuroscientist specializing in color vision. This person knows all the objective science about how color is experienced. However: she is color-blind (achromatopsia). She knows everything there is to know about red, but she doesn't *know* "red."

This subjective, experiential, hard-to-define but undeniable aspect is called *qualia*. This is one of the hot topics in neuroscience and may require a radical change in the scientific approach. As discussed in II.5: "Causality: A Map of the World as a Machine," science has always sought an objective, "third-person" perspective and finds an "autopsychological" approach impossible.

At its core, it is the classic subject/object polarity that arises in quantum mechanics or chaos theory and is central to Zen thought (Suzuki, 1969) (see IV.1: "Resonances from the East").

The "Second Derivative": A.R. Damasio's Hypothesis

In 1999, the neuroscientist Antonio R. Damasio proposed an architecture to explain the phenomenon of consciousness (Damasio, 1999). According to him, there are several layers in the genesis of consciousness.

1. A first layer (**proto-self**), which would gather afferences related to basic survival aspects (autonomic, proprioceptive, hormonal systems). This layer would be responsible for emotions (in the strictly physicochemical sense) not yet conscious. Structurally, it would be related

to brainstem nuclei (including the ascending reticular system), the hypothalamic area, and the somatosensory cortices. Its dysfunction would lead to a coma (loss of consciousness, including both senses of the term: cognitive and alertness).

2. A second layer (**core consciousness**) would gather afferences from both the proto-self and sensory (or memory) afferents, as well as, crucially, from the changes these sensory stimuli produce in the proto-self. It should be noted that afferences can also be non-sensory: memories or imagination can also produce changes in the proto-self, "activating" core consciousness. The second part of its function is to modulate these sensory afferents: deciding which ones "take center stage" (attention mechanism) or which can remain peripheral (and maintain automatic behaviors). In this layer, consciousness begins, but it's a point-like, pulsatile consciousness, constantly updated. Core consciousness would be organized over structures of the thalamus, the cingulate cortex, and the prefrontal cortex. Its dysfunction would lead to a condition where consciousness-alertness is maintained, but not consciousness-knowledge. An example would be akinetic mutism[1] or epileptic automatisms.[2]
3. The pulses of core consciousness would be stored in a memory system, serving as the basis for autobiographical identity. The gradual storage and its subsequent manipulation constitute what Damasio calls **extended consciousness**, which would allow learning, could modulate core consciousness, and serve as the basis for what we understand by personality. This function would be much more distributed across associative cortical areas. Its dysfunction leads to conditions like transient global amnesia,[3] anosognosia,[4] or asomatognosia.[5]

It's hard not to see a parallel between this architecture and the Freudian model. Damasio's proto-self resembles the Freudian Id, preconscious, closely linked to primary and survival emotions. The Freudian Ego is not far from core consciousness, generating individuality and a sense of time. Perhaps not coincidentally, the structures Damasio associates with core consciousness are part of the Default Mode Network (Damasio, 2010), which has also been associated with the Freudian Ego (Carhart-Harris & Friston, 2010). Finally, extended consciousness, with its cultural influence and acquisition through experience, resembles Freud's Superego.

Consciousness as a Map

To Damasio, consciousness would be a kind of Map of the effects produced by stimuli (sensory, proprioceptive, mnemonic) on the proto-self. In a way, consciousness would be information processing itself. To endure over time, this point-like consciousness can be stored in a memory system (autobiographical identity), in a way that evokes the function of writing for oral storytelling.

As with any process of jumping from Territory to Map, it is a selective process that omits and loses some aspects. In return, it gains the capacity to accumulate, retain over time, generalize (apply to similar but not identical situations), link (reasoning), and transmit (teaching/learning). This confers enormous evolutionary advantages: it increases predictive capacity, thus improving flexibility and adaptability to changing conditions while, at the same time, it allows for learning (which would be akin to software updating without changing the hardware). As mentioned in section I.3: "The Grand Journey," this accelerates the pace of evolution because it allows for changing the phenotype without modifying the genotype (hardware).

Footnotes

1. A condition in which the patient is awake with his vital functions preserved but is unable to speak or move.
2. During some epileptic seizures, patients remain awake and can perform some automatic actions (i.e. walking or drinking from a glass), but are not aware of them and do not remember them afterward.
3. Condition in which the patient loses the ability to form new memories (anterograde amnesia) but retains other vital functions (including remote memory). This condition usually resolves spontaneously.
4. The patient is unaware of his deficit. For example, he might have paralysis on the left side of his body. When asked what's wrong, he would answer that nothing is wrong. When asked to lift his left hand, he would use his right hand to grab and lift the left one.
5. The patient is unaware of part of his body. For example, he might be unaware of an entire half of his body, and when trying to pass through a narrow doorway, he would consistently bump into the door frame.

II. Border Posts

The Map/Territory polarity is ubiquitous. As soon as one becomes aware of this dichotomy, it's impossible not to find it everywhere. This section will discuss some specific fields that have developed very much in relation to this frontier, or in which this duality is especially evident. The list is obviously not exhaustive, and I hope that the reader will take pleasure in seeking and finding this duality in almost any aspect of everyday life.

II.1: Time

Time-Territory, Time-Map

The Map/Territory dichotomy is well reflected in our conception(s) of Time, and was probably already considered in classic Greece with the dual naming of *kairos* and *chronos*.

Territory-Time

There exists a lived, experiential Territory-Time, often represented as a river in which we are immersed. It's an absolute, immutable, asymmetrical, and irreversible time. Although conceptually difficult to define, it is intuitively evident. It is the Time involved in Literature, aging and death. However, as intuitively undeniable as it may be, it poses significant problems. Our conception of Territory-Time has evolved over millennia, during which we have moved in spaces and at speeds where there are no significant differences between classical physics and quantum or relativistic physics. This has inculcated a conception of time as a uniform constant, absolute, independent of space and speed.

However, at the beginning of the 20th century, Relativistic Physics dismantled this conception, demonstrating that time does not "flow at the same speed" everywhere (Einstein & Infeld, 1938). Even knowing this, for some years, we could live ignoring these facts without facing real problems. But our technology can no longer omit these details (for example, the geolocation systems we use (GPS, mobile apps) introduce this factor to improve location). Similarly, Quantum Physics contests the idea of homogeneous time and postulates a "granular" time, with an indivisible "unit of time" (on the order of 10^{-44} seconds), below which it is impossible to talk about time.

In short, we are in a similar situation to that of Renaissance man with the spherical Earth: he could "know it intellectually" without truly believing it, and it did not affect his life. This is no longer sustainable. How long can we maintain the fallacy of absolute time?

Map-Time

There's also a Map-Time, which is the measurable and calculable time. Although it seems a simple abstraction of Territory-Map, it requires a prior manipulation that denaturalizes and profoundly changes it. Essentially, this Map-Time needs to be projected and turned into a homogeneous space (Bergson, 1927), where this sequentiality can be treated "as a whole" (and it is no coincidence that the brain networks that handle time perception are closely linked to spatial management (Buonomano, 2018)). Treated in this way, Time acquires new characteristics: it becomes homogeneous and symmetrical, can be sped up or slowed down, and can even be reversed.

This "denaturalization" offers huge possibilities in return, most notably the capacity for abstract handling: it can be introduced into calculations and allow for predictions. This is the "time" dimension that is at the origin of the Calculus invented (or discovered?) by Newton and has fueled the explosive development of Physics and Technology.

For many centuries, there were no significant frictions between Time-Map and Time-Territory. Each locality had "its" time, generally marked by the church bells.

The development of great journeys modified the situation. Maritime navigation needed a precise cartography (longitude and latitude) to be able to locate territories. The problem of latitude was quickly solved (the origin latitude (0º: Equator) is an empirical fact, and soon it could be measured by astronomical systems). The problem of longitude proved much more arduous. Fundamentally, measuring the longitude was nothing more than estimating the time difference between one position and the meridian considered as a reference. Therefore, solving this problem became a matter of the technology of chronometry (Sobel, 2006).

A qualitatively new step has been the digitalization of this Map-Time. The classic example is the Fast Fourier Transform (FFT), a widely used technique employed to digitize image or sound files. Referring specifically to sound files, it is striking how time can be "captured" in space, and manipulated (transmitted, compressed, expanded...) arbitrarily.

The FFT (see II.7: "Fourier Transform") is based on the capacity to represent any shape by a sum of wave trains with certain frequencies, amplitudes, and phases. Thus, a time series can be described "instantaneously" by a set of frequencies and amplitudes. In engineering, this is called moving from "time mode" to "frequency mode," and fundamentally, it involves collapsing a time series into space. Of course, "experiential" time disappears along the way. Ideally, if this collapse were perfect, we would be in the situation of the God mentioned by Pierre-Simon Laplace (Wikipedia, 2017, Pierre-Simon Laplace): "...for such an intellect nothing would be uncertain, and the future, just like the past, would be present before its eyes."

However, there is always a gap. A "complete" description of the time series would require an infinite set of frequencies. In practice, only a finite set of frequencies can be used, and therefore, the FFT necessarily loses details. As we will see in several other examples (calculus, fractals, etc.), the concept of infinity is a bridge (or a barrier) between the Map and the Territory.

As with all examples of the leap from Territory to Map, the digitization of an analog reality implies a loss of detail (in exchange for immense power). This gap between Map and Territory (which we will see every time the leap occurs) can be made as small as desired. For the purposes of sound files, the difference may be imperceptible to the human ear (indeed, digitization techniques have significantly improved the quality of most sound recordings), but this gap is essentially insurmountable.

Presentism/Eternalism

These difficulties in the concept of Time are reflected in two distinct ways of conceiving it (Buonomano, 2018; Rovelli, 2019).

The presentist school considers that only the present is real. The past is maintained only in memory, and the future is just hypothetical and undetermined. This conception melds conveniently with our sensation of time as a flow that passes through us, while we only exist in the present.

In contrast, eternalism argues that the present (the "now") is similar to "here" and is only relevant to the observer, but cannot become an absolute category. One of its formulations uses a Minkowski spacetime, with three spatial dimensions and one temporal dimension. By analogy, if we imagine a two-dimensional reality (living in a plane), we could make a kind of tube of sequential planes, so that each plane corresponds to a temporal moment, and the "time" dimension would be the axis of the tube (Fig. II.1.1):

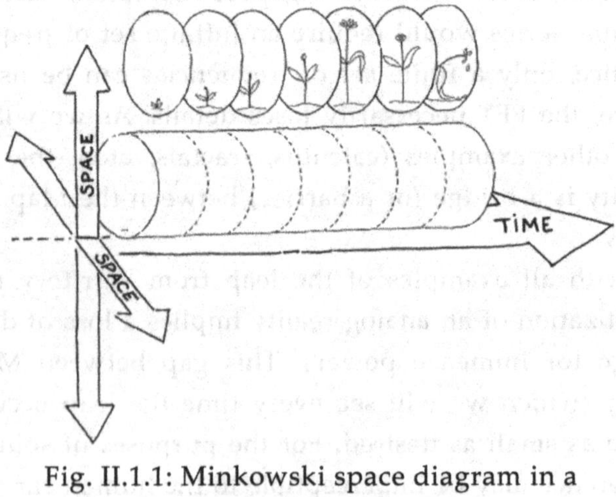

Fig. II.1.1: Minkowski space diagram in a two-dimensional reality.

A Minkowski spacetime would be the extension of this model to a three-dimensional space with time as the fourth dimension.

The presentism/eternalism polarity also tunes with the classic distinction between episodic and semantic memories. Episodic memories evoke specific moments, often with a strong emotional charge (e.g., the death of a loved one, a special romantic relationship, etc.). In contrast, semantic memories are "Map-information" (e.g., what is the capital of Russia, or what is a rhinoceros). Episodic memories have a temporal stamp (and among several, we can order which happened first). Semantic memories do not have this temporal stamp, and we probably cannot remember when we learned the capital of Russia unless we refer to an episodic memory (e.g., geography classes at school). Yet, we need semantic memories to record episodic memories (to remember the scare of seeing a tiger, you need to know what a tiger is).

In summary, we can consider a presentist, Territory-time, supported by episodic memories, versus an eternalist, Map-time, supported by semantic memories.

The Problem of the Observer

It's challenging to overcome the naive impression that the time we perceive is a faithful reflection of an "objective," external, and constant variable. This very intuitive idea of time as a path or river is one of the clearest examples of the "somatic metaphors" in which we are trapped (Lakoff & Johnson, 1999, 2013) and which are so difficult for us to overcome. When these metaphors are really successful they become invisible, and we come to consider them a faithful representation of reality. Thus, we understand our eyes as windows through which we see Reality. In the case of time, we consider it an objective reality, external to the individual, that flows (the metaphor of the river or the path). This decisively conditions our way of relating to it and we forget this metaphorical component (D. Hoffman, 2015; D.D. Hoffman, 2016).

And yet, this view is already untenable. On one hand, as mentioned before, quantum physics and relativity have produced evidence that is incompatible with this "common sense conception."

On the other hand, there is plenty of data demonstrating that perceived time is a deeply "edited" variable. There are numerous examples of how the perception of time depends on spatial variables. For instance, it is known that the perception of the time separating two phenomena is influenced by the physical distance between the phenomena (kappa effect, Huang & Jones, 1982), and there is a tendency to represent time on a horizontal line (STEARC effect, Ishihara et al., 2008). It can be shown that our minds "erase" the time in which eyes blink or have saccadic movements (Burr et al., 1994) (which can extend up to an hour a day!); there is a "temporal integration window" in which the mind "glues" as simultaneous two temporally offset stimuli, but which "we know" are simultaneous (e.g., in a concert the sound of the drums and the image of when they are struck, or the image and the voice of many videos), and so on.

In summary, the idea of an "objective" perception of time does not hold up from the point of view of physics nor from the perspective of neuroscience.

From a Darwinian point of view, it is reasonable to think that our perception of time is an evolutionarily very successful mechanism. It allows us to learn from the past and predict future situations, and it seems indisputable that it improves the "fitness" of the species. Given the context of spaces and speeds in which we have developed, we have internalized a "Newtonian" physics, imprecise, but much simpler, and sufficient for the world in which we moved. But that model is already unsustainable, and it is not obvious how to harmonize the kairos time of the mind with the chronos time of physics. Brian Greene expresses this difficulty well:

> *Is science unable to grasp a fundamental quality of time that the mind embraces as readily as the lungs take in air, or does the human mind impose on time a quality of its own making, one that is artificial and that hence does not show up in the laws of physics? If you were to ask me this question during the working day, I'd side with the latter perspective, but by nightfall, when critical thought eases into the ordinary routines of life, it's hard to maintain full resistance to the former viewpoint.* (Greene, 2004)

Naturally, the problem of time is intimately linked with consciousness. Even if we accept Damasio's model of a pulsatile proto-consciousness (Damasio, 1999), full consciousness (Damasio's extended consciousness) requires a memory that guarantees continuity, and therefore is already automatically immersed in time. And in this sense, it's difficult not to suspect that the Default Mode Network (see Introducing the DMN in I.4: "Mind/Map versus Brain/Territory") plays a crucial role in these functions of temporal perception:

- it is a system intimately related to psychological "time travel" (Buckner & Carroll, 2007) (e.g.: What was I like as a child? What will I be like as an old person?)
- its connectivity increases from childhood to adulthood (Fair et al., 2008; Gao et al., 2009) (just as the tendency to time travel increases)
- in certain situations (e.g., with the use of psychedelics) its activation is associated with a significant decrease in brain entropy levels (Carhart-Harris et al., 2014). And although this concept of entropy refers to informational entropy, we cannot forget its connection with the idea of

thermodynamic entropy (see II.2: "Entropy(ies)"), nor the fact that thermodynamic entropy is ultimately the "arrow of time": the ultimate (and only) reason why time seems irreversible to us.

It is possible that new Artificial Intelligences will be developed that are not restrained by our "somatic metaphors" and may handle the idea of Time in ways we cannot imagine. Could this be one of the points of disconnection between AIs and consciousness as we know it?

II.2: Entropy(/ies)

Although it may seem conceptually distant, a topic intimately linked to the border between the Map and the Territory is the idea of entropy.

Thermodynamics

The concept of entropy emerged during the Industrial Revolution, with the rise of the steam engine and the development of Thermodynamics. It was observed that when a temperature gradient exists, work can be produced. When this gradient disappears, the energy persists (it is constant in a closed system), but it is inaccessible. The initial concept of entropy referred to the characteristics of this non-usable energy. This is the idea underlying the second law of Thermodynamics: the entropy of a closed system can only increase. The corollary is the thermic death of the universe: over time, complete thermal equilibrium will be reached, and all changes will cease. If Aristotle's idea of time as a measure of changes is accepted, this means that time will stop.

Statistical Mechanics

Statistical Mechanics tried a more precise approach to the problem. Assuming that the temperature of a gas was the average kinetic energy of its randomly moving molecules, it was easy to understand that collisions between molecules would end up homogenizing the velocities, and therefore the temperature. Entropy became a measure of the probability of the set of microstates of a system. Formally,

$$S = -\sum p_i log_2 p_i$$

where S is the entropy, p_i is the probability of each microstate, and log_2 is the base-2 logarithm. The maximum entropy is reached when equilibrium is achieved and all microstates are equal (i.e., indistinguishable to an observer). Imagine a compartment with gas that is initially hotter on one side than the other. Over time, the random movement of the molecules will make it unlikely to maintain an "organized" configuration (fast molecules on one side, slower on the other), and the probability of a "disorganized" structure (fast and slow molecules mixed) will increase. Thus, entropy came to be measured as "disorder," assuming that disorder tends to increase, and it is extremely unlikely that molecules will "organize" themselves spontaneously.

This also had enormous utility in other fields. In conventional physics ("Map"), the "time" variable is homogeneous, and does not have a predetermined direction. The equations work just as well with increasing as with reversing time. However, in real life ("Territory"), time is not reversible: it is almost the very definition of irreversibility. Entropy came to solve this paradox by introducing an irreversible "arrow of time": although any random movement may have points that "reorder" spontaneously, overall, this probability is ridiculously small. When statistical mechanics comes into play, entropy can only increase, and this determines an irreversible temporal direction.

However, this seemingly objective conception of entropy is misleading. What constitutes an "ordered" structure depends on the observer's idea of "order." The sequence "6, 9, 3, 1, 4, 7, 2…" may seem random, or it may be a highly ordered structure (decimal expansion of the logarithm of 2). The very difference between "macro" and "micro" states presupposes the existence of an observer capable of distinguishing them.

Therefore, entropy is not an isolated and self-sufficient concept: it inexorably requires an observer. An atlas or its equivalent weight in paper have the same entropy if we want to burn it for heat, but obviously not if we consider the order it contains (in the distribution of the ink dots). Moreover, entropy crucially depends on the limits of the system being considered. Although the second law of Thermodynamics is irrefutable for closed systems (ultimately, hypothetical), it can be violated in open systems. Thus, one of the characteristics of living systems is precisely the ability to produce order (see below), almost always residing far away from equilibrium. Membranes, which are one of the essential characteristics of living beings, function as true "Maxwell's demons" (Wikipedia: Maxwell's Demon), capable of selectively allowing certain molecules to pass and thus building a gradient (and consequently reducing local entropy).

Information Theory

Information Theory provides another approach to the concept of entropy. Shannon distinguished between the amount of information and the idea of meaning, proposing that the amount of information in a message can be calculated by analyzing the uncertainty it contains (and eliminates upon receipt). For instance, in the series "1,1,1,1,1,1..." it is trivial to predict the next term, thus carrying very little information (very little uncertainty is destroyed). With a coin toss (heads or tails), the chance of guessing correctly is ½. With dice, it's 1/6 (six faces), and for a completely random series of digits, it's 1/10 (10 possible digits: 0, 1, 2, ... 9). For a random series of letters, it's 1/26 (a, b, c, ... z). Shannon introduced a measure of the quantity of information, H, which he also termed entropy, and its formula is similar to that of thermodynamic entropy S

$$H = -\sum p_i log_2 p_i$$

where p_i is the probability of each possibility.

Applying this equation to the previous examples, the entropy associated with each term in the series would be:

Table II.2.1

Type of random series	Possible outcomes	Entropy
series 1,1,1,1,1...	1	0
coin toss (heads/tails)	2	1
dice (1–6)	6	2.58
series of digits (0–9)	10	3.32
series of letters (a–z)	26	4.70

Entropy as the Distance between Map and Territory

It should be emphasized that Shannon's idea of information is not related to meaning, but to the elimination of uncertainty. Therefore, what it truly measures is not how much we find out (in the sense of knowledge), but how much we don't know (in the sense of unpredictability). Essentially, entropy reflects the gap that separates (our) Map from (the) Territory. A high-entropy system speaks of a model poorly adjusted to reality, requiring frequent "measurements" to "stick to the terrain." A low-entropy system tells us about a model that *a priori* fits its territory.

Entropy in Biological Systems

The metaphorical equation of entropy with disorder (e.g., "life is an example of continuous entropy reduction") can be misleading. Kate Jeffery et al. (Jeffery et al., 2019) propose several illustrative models. For instance, consider a sealed box with multiple balls randomly distributed, moving at different speeds. If the collisions were perfectly elastic, the system would continue indefinitely. Since this is not possible, each collision results in a reduction of speed and an increase in temperature. In the final equilibrium situation, there will be an increase in "macro" order (all balls stationary on the floor, instead of scattered through space) but also an increase in the system's entropy (increase in the temperature of the balls and the walls). In summary, there is an increase in "order" in exchange for a transfer of entropy from the "macro" state (balls in the room) to the "micro" state (movement of the molecules in the balls and walls).

This is the norm in processes of increasing order in Biology. Metabolic cycles and the processes of creating structures channel entropy toward the microstates (or to the outside). Consequently, the system's overall entropy increases. This is only possible far from thermodynamic equilibrium, and with access to an energy source (in the case of the balls, gravity; in biology, the sun or energy stored in chemical bonds).

In this sense, it's not so much that biology produces order in exchange for energy. It would be more accurate to say that it has found a way to increase global entropy (following the slope of the second principle of Thermodynamics) by transferring entropy from the macro level to the micro level.

This "entropic slide" goes through successive meta-stable states: local minima of entropy that require an extra energy input to "overcome the step" and continue down the (increasing) entropic slope. An example would be a pile of firewood: it is in a meta-stable situation, but with an energy input (a match), it can burn and continue increasing entropy. If we consider the phase space of a biological system (i.e. the set of all possible states of the system), Jeffery et al. (Jeffery et al., 2019) propose a sponge-like image (Fig. II.2.1), with the meta-stable states represented by bubbles connected by channels that would allow the "entropic flow": a mechanism to increase complexity at the expense of transferring entropy from the "macro" state to the "micro" state.

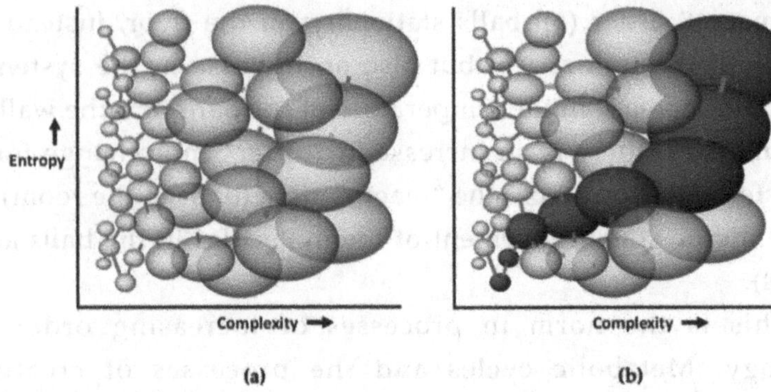

Fig. II.2.1: (a): Simplified diagram of the phase space of a biological system, with multiple meta-stable regions ("bubbles") connected by channels. (b): the system moves from one bubble to another (shaded bubbles) and will find it easier to enter large bubbles and more difficult to exit them. Therefore, the tendency will be towards greater complexity. Figure obtained from Jeffery, K., Pollack, R., & Rovelli, C. (2019). On the Statistical Mechanics of Life: Schrödinger revisited. *Entropy* 2019;21,1211, with a Creative Commons license.

The increase in complexity would be associated with increasingly larger (more stable) bubbles, and consequently the probability that the transition from one state to another occurs due to stochastic fluctuations becomes ever smaller. Instead, structural mechanisms that facilitate this transition are developed: essentially, enzymatic systems. These molecules, with a very specific structure, allow for channeling the "entropic flow," greatly increasing the probability of exiting the meta-stable state.

The concept of complexity, although intuitive, is not easily formalizable. A first approximation would be the presence of patterns or symmetries: the idea that, by knowing part of the model, information about the other part is obtained. However, it is important to conceive of complexity not only in a structural sense but also in a dynamic/temporal sense. In general, complex systems can only be understood by considering this temporal aspect (e.g., planetary orbits can only be understood dynamically). And a clear example of "temporal symmetry" is the reproduction of models. Naturally, this is one of the central characteristics of living beings. In a phrase attributed to Francis Crick, *"life is the way DNA has of making more DNA."* Within this process of producing dynamic symmetries, a crucial step is the ability to create internal cognitive models that attempt to reflect the outside: this is the basis of the learning/knowledge processes. Another corollary of this dynamic vision is the necessity of death: only through the death of individuals (and the reuse of their material) can new paths of complexity be explored. Naturally, some paths (or perhaps, eventually, all) end in dead ends (and ultimately, extinction).

On this path of complexity, Jeffery et al. propose several particularly crucial steps:

1. The transition from isolated replicating molecules to populations of compartmentalized molecules (i.e., the appearance of the cell).
2. The transition from unbound replicants to chromosomes.
3. The shift from RNA acting as both gene and enzyme to DNA and proteins (discrimination of genetic coding and protein assembly).
4. The emergence of photosynthesis (and consequent production of oxygen and mass extinctions).
5. The transition from prokaryotes to eukaryotes.
6. The appearance of genetic regulatory systems, which allowed for cellular differentiation and the development of complex multicellular life forms.
7. The transition from asexual clones to sexual reproduction, with all the combinatorial capacity this implies.
8. The shift from solitary individuals to colonies (e.g., social insects).

9. The development of neurons, which enable coordination of actions across cellular conglomerates and organizing specialized behaviors (e.g., muscle contraction).
10. Synaptic plasticity (the ability to modulate the intensity of connection between neurons). This makes it possible to save previous models and is the origin of memory, allowing organisms to remember the past in order to foresee the future.
11. The emergence of internal cognitive representations. This allows simulating reality and optimizing predictions.
12. The development of language and symbolic reasoning. This permits the transmission of accumulated information both over time and across space.

As discussed in I.3: "The Great Journey," it's not hard to see the progressive expansion of information/Map in this evolutionary development. The very transition between metastable states (the channels that communicate the "bubbles") is crucially dependent on enzymatic systems, that is, molecular configurations: another data/Map.

Disease as a Loss of Complexity

Entropy can therefore be interpreted both as "amount of information" or as "uncertainty," and this ambiguity heavily influences the techniques used to analyze complexity in biological systems.

Many biological systems are characterized by integrating multiple feedback and feedforward mechanisms, and exhibiting an unpredictable behavior typical of deterministic chaos (Goldberger et al., 1990). One of the earliest signs of dysfunction in these systems is a "simplification" (loss of complexity) in the structure or behavior of these systems, characterized by the appearance of repetitive, rhythmic, or predictable behaviors (senile tremor in Neurology, loss of biodiversity in an ecosystem in Ecology, decrease in heart rate variability in Cardiology, etc.). It is assumed that the loss of complexity is a consequence of structural deterioration or a decrease in processing capacity, which causes a worsening of fine control, thus requiring larger oscillations to sense changes and trigger appropriate control mechanisms.

The classic tool to study these changes in system complexity is entropy analysis (in any of its various techniques) (Vargas et al., 2015) (see II.8: "Renormalization and Complexity"). Generally, entropy loss is a very early and sensitive indicator of decline.

This would be quite counterintuitive if entropy is considered as a measure of disorder (concept from Statistical Mechanics), but not with the idea of entropy as a quantity of information (Information Theory): a healthy system ("rich in information") would display high entropy, which would decrease as the system deteriorates.

In recent years, complexity analysis (or, more precisely, the analysis of its loss) has become a very sensitive tool for detecting early stages of certain diseases in many fields (Cardiology, Endocrinology, Neurology…) (Goldberger, 1996; Goldberger et al., 1990; Lipsitz & Goldberger, 1992; Varela et al., 2010; Vargas et al., 2015).

However, there is the paradoxical situation where, in some cases, dysfunction is characterized by an increase (and not a decrease) in entropy (Vargas et al., 2015). This could be explained by a loss of efficacy of the control mechanisms, and therefore an increase in truly random factors. It seems then that the entropy we measure (uncertainty-entropy) inseparably includes both the fine control mechanisms that we are unable to measure ("health") and the random aspects ("loss of control"). This entropy/gap-between-Map-and-Territory (a kind of Janus-faced entity, in which it is impossible to dissect its two components) is at once one of the most powerful tools for analyzing some physiological systems, and a barrier that prevents their modelling.

II.3: Differential Calculus; Concept of Infinity

An example of the boundary between Map and Territory is calculus (Acheson, 2017) developed (probably independently) by Newton and Leibniz in the second half of the 17th century. Broadly speaking, it involved approximating the measurement of the area under a curve by evaluating the areas of underlying rectangles.

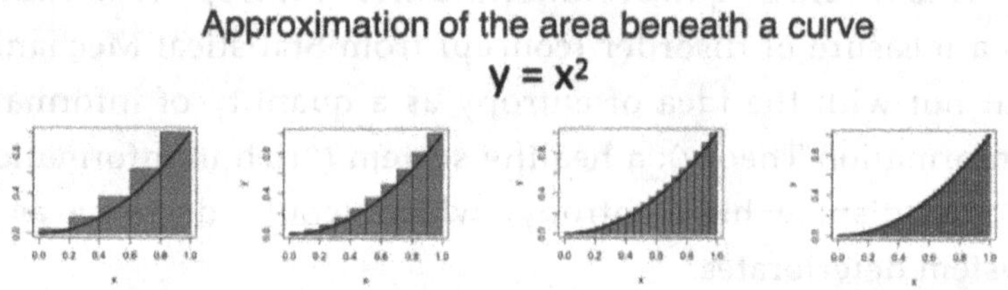

Fig. II.3.1: Approximation of the area beneath a curve $y = x^2$

The difference between the actual area under the curve ("Territory") and the approximation based on summing the areas of the rectangles ("Map") will decrease (and therefore the approximation will be better) the smaller the "steps" of x are, and in the end, the problem is reduced to obtaining the slope of the curve at each point (this is precisely the concept of the derivative). The mathematical formulation would be:

$$\frac{dy}{dx} = \lim_{\delta x \to 0} \frac{\delta y}{\delta x}$$

This deals with a typical example of the leap from the Territory to the Map: as usual, the process involves quantizing a continuous variable (in this case δx). This requires a "simplification" that involves omitting details considered irrelevant. In return, an "autonomous" Map-object is obtained, which can be manipulated, stored, communicated, and introduced into complex reasoning.

The utility of calculus was evident from the beginning, but from a conceptual standpoint, it was considered not very rigorous, and its theoretical basis was much debated among mathematicians. An example of the skepticism it generated was the essay *The Analyst: a Discourse Addressed to an Infidel Mathematician*, published in 1734 by Bishop George Berkeley (Berkeley, 1734). Regardless of its religious objective, the text presents significant arguments against the justification of infinitesimal calculus. In its original formulation, for example for the curve $y=x^2$, the increment Δy calculated for an increment Δx would be:

$$\Delta y = \left(x+\Delta x\right)^2 - x^2 = 2x\Delta x + (\Delta x)^2$$

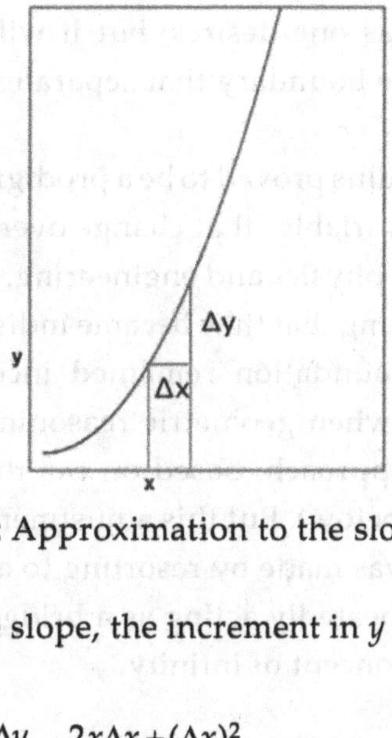

Fig. II.3.2: Approximation to the slope of $y=x^2$

Then, to obtain the slope, the increment in y was divided by the increment in x:

$$\frac{\Delta y}{\Delta x} = \frac{2x\Delta x + (\Delta x)^2}{\Delta x} = 2x + \Delta x$$

Finally, in the last step, Δx was "minimized until it disappeared," and therefore the derivative of the curve [$y=x^2$] was 2x. But Bishop Berkeley asked: Is $\Delta x = 0$ or not? If $\Delta x \neq 0$, it cannot be eliminated. On the other hand, if $\Delta x = 0$ the fraction is incorrect, 0 in the denominator was not acceptable.

Leibniz himself, co-inventor of infinitesimal calculus, had his reservations about the method:

> *It is true that for me, infinities are not a whole and the infinitesimally small is not a size. My metaphysics banishes them from its lands (...). The part that I have used in the calculus of infinitesimals has not endeared them to me beyond common sense itself...* (Leibniz, 1857)

This point about "evanescent quantities" is similar to the difference between the real area under the curve $y=x^2$ and the approximation by summing the area of the rectangles: it can be made as small as one desires, but it will never disappear. This is precisely the boundary that separates the Map from the Territory.

In any case, calculus proved to be a prodigious tool: it allowed for the analysis of variables that change over time and signified a true revolution in physics and engineering. Its practical results were so overwhelming that they became indisputable. However, its mathematical foundation remained inconsistent until the mid-19th century, when geometric reasoning was replaced by a more algebraic approach, based on the theory of limits and infinite series (see below). But this adjustment between the Map and the Territory was made by resorting to a turbid instrument that we will see repeatedly acting as a bridge between the two: the mathematical concept of infinity.

Infinity

> *The attempt to capture infinity with the nets of language, through definitions or algorithms, by analogies or homonyms, makes it evident that if the task seems vain, it is no less so than trying to avoid the subject.* (Lévy, 1987)

The concept of infinity is inevitable when discussing the Map/ Territory polarity. It is a gap (or a bridge) that connects or separates these two forms of reality. As we will see, it is not only an unavoidable concept. It is also a central tool upon which some of the most powerful scientific instruments rest (differential calculus, fractal geometry, Fourier transform, etc.). Furthermore, intimately linked with the problem of the infinitely large is the problem of the infinitely small: can space (or time) be divided indefinitely? Or is there a final point, definitively indivisible? This issue raises the question of the nature (continuous or discrete) of reality. This question (present since classical Greece) continues to resonate in contemporary physics (wave-particle duality, string theory, etc.).

Historical Background

The idea of infinity must be very ancient: it seems an inevitable conclusion as soon as one starts to count: given any number, it is always possible to add one more unit, indefinitely. The idea of an infinite number of natural numbers seems inevitable.

However, the concept is fraught with problems and paradoxes (e.g., all numbers are either even or odd, is infinity even or odd? If we subtract a quantity from the total of infinity, will it always remain the same, regardless of how much is subtracted? Etc.).

As Tony Lévy (Lévy, 1987) points out, the "cultivators" of infinity have changed over the centuries. In classical Greece, this was a topic mainly addressed by philosophers (and included in the study of what we would now call Physics). In the Middle Ages, it was the domain of religions and referred mainly to the attributes of God. Only from the 19th century did it become a legitimate object of Mathematics.

From the first moment, it was a controversial concept. Its elusiveness was justified in very different ways. Aristotle accepted a "potential infinite" (for example, he considered that time was infinite), but flatly rejected the existence of an actual infinite. Archytas of Tarentum supported the spatial infinite arguing that, if one stands at the supposed limit and extends an arm, what does one find? (naturally, this ignites a whole controversy about the vacuum). Some authors stated that infinity is immeasurable, and therefore it makes no sense to talk about its size or compare it with other quantities (Nicolas Oresme), or that the concept of infinity is just a proof of our cognitive limitation, and that such elusiveness would not be such for God (Augustine of Hippo), or that the idea of an infinite quantity is imaginary and does not respond to an objective reality (Spinoza).

Lévy describes how in the three Western monotheistic religions (Christianity, Islam, and Judaism) there was a clash between two ways of combining philosophy and religion, and in all of them the problem of infinity occupied an important place. On the one hand, there were the philosopher-thinkers, who at no point question faith, but consider that philosophy will never contradict dogma (Thomas Aquinas: *"Reason, which exists in the mind of God and governs the whole universe"*) and can be a powerful tool to spread it. On the other hand, some theologians prioritized faith and teachings above the supposed philosophical reasoning and feared the noise generated by paradoxes and contradictions. In the Christian world, this led to the edict in 1277 forbidding the teaching of 219 articles that are supposed to collect the body of Aristotelian philosophy. In the Islamic world, one of the key figures of Sunnism, Al-Ghazali, published in the 19th century *The Incoherence of the Philosophers*, with very serious accusations of infidelity towards the *falasifa* (philosophers). In 1305, the Jewish authorities of Barcelona prohibited "philosophical studies."[1] A significant part of all these controversies revolves around the infinite. Many thinkers of classical influence rejected the idea of an actual infinite (as opposed to potential), and in any case attributed to the infinite a certain defective character, impossible to traverse, not bounded. Conversely, for most theologians it is a sublime characteristic, which manifests the power and magnificence of God.

Probably one of the great sources of paradoxes and controversies around the infinite refers to the impossibility of bounding it, of setting limits. Therefore, an important step was to take the concept out of the physical space-time realm, and "arithmetize" it, transferring it to the realm of arithmetic. There it was seen that many series tended to infinity, e.g., the "harmonic series":

$$1+\frac{1}{2}+\frac{1}{3}+\frac{1}{4}+\ldots=\infty$$

but there were other series, also infinite, that converged to a calculable number, for example:

$$\frac{1}{2^0}+\frac{1}{2^1}+\frac{1}{2^2}+\frac{1}{2^3}+\ldots..=2$$

As an example of the previous series, one could imagine somebody standing in the center of a circle with a diameter of 2 meters. He tries to leave the circle, but with each step, he can only cover half the distance that separates him from the edge (the first step would be 1 meter ($1/2^0$), the next 50 cm ($1/2^1$), etc.). He would need an infinite number of steps to reach the edge.

This process of arithmetizing infinity by removing it from the spatial-temporal realm provided a solid foundation for infinitesimal calculus, moving it from the domain of "engineering" to that of pure mathematics.

Relativistic Physics also twisted the relationship between infinity and limits. Some relativistic models based on Riemannian geometry postulate a finite but unlimited Universe (Zalta et al., 2021). This idea, as counterintuitive as it may seem, can be approximated by thinking about a creature that only considers one dimension, and whom we place on a circumference. The creature could progress indefinitely without encountering limits, although from a 2D perspective their universe is finite. Similarly, a creature living in 2D and standing on the surface of a sphere could travel it indefinitely (without limits), although in 3D its surface is finite. If we, "3D beings," lived in a 4D Universe, we might find ourselves in a similar situation: we could travel in a "straight line (3D)" to arrive at the same spatial point from where we started.

The historical discourse on the concept of infinity questions what can be considered a demonstration (the moment when the listener "sees": grasps and accepts the argument). The argument of Archytas of Tarentum, previously cited, would not convince anyone familiar with relativistic physics. One of the central arguments in the history of infinity is the issue of the infinitude of individual souls: if the world were eternal (as Aristotle thought), there would be infinite souls, contradicting the nonexistence of an actual infinite. Probably no one would take this argument seriously at present, but for scholasticism, it was one of the hot points of the concept of infinity.

Some of the paradoxes used to demonstrate that the idea of infinity is self-contradictory have become central pillars in the current concept of infinity. One of the classic issues derived from the problem of the whole and the parts. Galileo pointed out the following paradox: if we select any number of successive natural numbers, there would always be many fewer square numbers than numbers in total; however, in the series of all natural numbers, one could assign a square number to each number ($n \rightarrow n^2$), which would demonstrate that there are as many square numbers as numbers in total. Currently, the definition of infinity given by Dedekind is precisely a set in which a bijection (a one-to-one relationship) can be established between part of the set and the entire set.

Similarly, one of the arguments against the concept of infinity was that if a finite quantity, as large as desired, is subtracted from an infinite set, the resulting set will still be infinite. This was considered a contradictory argument that questioned the very concept of infinity. Nevertheless, the entry of infinity as a proper subject of Mathematics arguably occurs at the end of the

19th century when Georg Cantor proved that, indeed, there are infinities of different sizes (see below).

Numbers: natural, integers, rationals, reals, imaginaries, transfinite...

The history of mathematics is strewn with numbers that were invented or discovered, which at first seem completely counterintuitive and gradually become obvious. The first humans who used numbers (naturals: 1, 2, 3 ...) for counting probably found the idea of negative numbers completely absurd (but eventually integers were consolidated: 1, -1, 2, -2 ...). Then came rational numbers (1/2, 3/5 ...), which establish the relationship between natural numbers and were supposed to capture everything measurable. The demonstration that there exist irrational numbers (e.g., $\sqrt{2}$), which cannot be obtained as the quotient of two naturals (Appendix 1), was a dramatic crisis for the Pythagorean school. Subsequently came the real numbers, which include rationals, irrationals (e.g., $\sqrt{2}$) and transcendentals (numbers that cannot be obtained from a conventional equation $ax^n + bx^{(n-1)} + cx^{(n-2)} + dx^{(n-3)} + \ldots + px + q = 0$). Imaginary numbers are also described (multiples of $i = \sqrt{-1}$) (see II.6: "Fractals").

These steps have not been without difficulties. For example, it is hard to accept the existence of transcendental numbers, which by definition cannot be obtained from an equation, cannot be named (infinite decimals) and cannot be located on the number line (Chaitin, 2005). This generated a bitter controversy at the beginning of the 20th century between the formalist mathematicians and the intuitionist school. The former defended the individual existence of all real numbers (in a kind of Platonic world), regardless of whether anyone "was looking for them." Conversely, the intuitionists proposed that each point on the number line (each real number) represented a process, eventually unfinishable, that unfolded over time, somewhat in the spirit of Aristotle's potential infinity. Eventually, the formalist view prevailed, but this perspective presents significant drawbacks for handling time in Physics (Gisin, 2020).

In the second half of the 19th century, transfinites were introduced by Cantor (Zalta et al., 2021). Cantor demonstrated that not all infinities are equal, and that the infinity of real numbers exceeds the set (also infinite) of natural numbers (Appendix 2).

Cantor was deeply religious, and this probably influenced his differentiation of the transfinite numbers and an "Absolute Infinite," transcendent, which cannot be known by man. He was fully aware of the philosophical consequences of his approach (Dauben, 1977), and at a crucial time when the Catholic Church was positioning itself in relation to the scientific advances of the time (encyclical *Aeternis Patris* in 1879), Cantor maintained correspondence with important theologians (including Pope Leo XIII himself) defending his view of the infinite. Cantor proposes that both finite and transfinite numbers can be considered in two ways. On one hand, as well-defined and differentiated mental objects, capable of modifying thought itself. He called this intrasubjective or immanent reality. On the other hand, there was a transubjective or transient reality that expressed itself in physical phenomena. Arguably, the Map/Territory polarity resonates in this contrast.

Likewise, the introduction of different types of numbers (and especially the transfinites) and controversies over their "existence" evoke the Platonic approach to the world of Ideas.

In any case, all this evolution is nothing more than a progressive readjustment of our Maps to optimize our survival in the Territory.

Footnote

1. "From today, and for a period of fifty years, no member of our community shall study the 'Greek' works of science and metaphysics, either in their original language (Arabic) or translated (Hebrew) before reaching the age of twenty-

five years; no member of our community shall teach these disciplines to a student under the age of twenty-five, for fear that these studies might distract the student from the Torah of Israel, which is superior to these teachings (...). The study of Medicine has been excluded from this prohibition, although it is a natural science." (Baer, 1978)

Appendix 1: Proof that √2 is irrational by *reductio ad absurdum.*

If √2 were rational, it could be expressed as a fraction of mutually prime numbers (greatest common divisor = 1):

$$\sqrt{2} = a/b$$
$$2 = a^2/b^2$$
$$a^2 = 2 \cdot b^2$$

This would mean that "a" is even, and therefore expressible as:

$$a = 2 \cdot c$$
$$a^2 = 4 \cdot c^2 = 2 \cdot b^2$$
$$2 \cdot c^2 = b^2$$

From this reasoning, it follows that both "a" and "b" are even numbers and have 2 as a common divisor, contradicting the initial postulate that a/b is an irreducible fraction.

Appendix 2: Demonstration that the set (infinite) of real numbers is larger than the set (also infinite) of rational numbers.

The proof is based on the key idea of orderability. Essentially, the reasoning is as follows:

1. It is possible to order the set of rational numbers. For example, this could be done with a table with all the naturals

(1,2,3...) as headers for rows and columns. Each cell would be the rational resulting from the quotient: "row number/column number." This table would include all possible rationals (some of them repeated, but this does not invalidate the proof).

Then, the set of cells can be ordered. For example, considering each cell as [row number, column number], an ordering could be done:

1 ⇔ [1,1]
2 ⇔ [1,2]
3 ⇔ [2,1]
4 ⇔ [1,3]
5 ⇔ [2,2]
6 ⇔ [3,1]
7 ⇔ [1,4]

Etc...

Graphically (Fig. II.3.3):

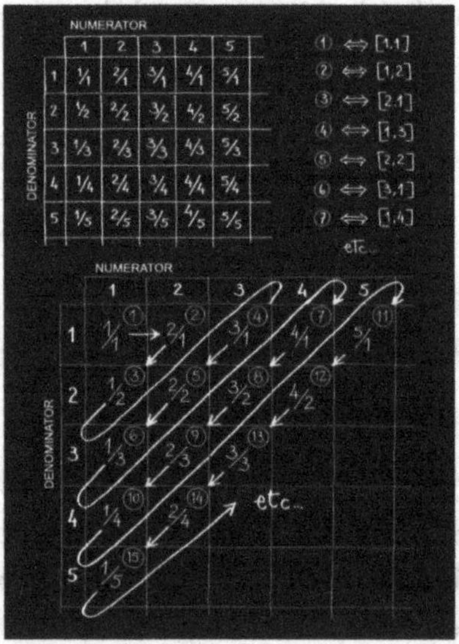

Fig. II.3.3: Ordering of the set of rationals

This allows the ordering of the set of rationals, and demonstrates that there is a bijection between natural numbers and rationals:

The infinity of the naturals is the same as the infinity of the rationals. Cantor called it ℵ0 (aleph zero).

2. It is not possible to order the real numbers. If we could make a list of real numbers (for example, between 0 and 1):

1 ⇔ 0.000.........

....

n ⇔ 0. $n_1\ n_2\ n_3\ n_4 \ldots n_n \ldots$

....

one could always obtain a real number that is not in that list. For example, it would suffice to change the first decimal from row 1, the second from row 2, the third from row 3, ... the nth from row n, ... etc. It can be seen that no number from the previous list is identical to the number we have produced (it will differ, at least, in the decimal place corresponding to its (hypothetically possible) order row).

This demonstration by *reductio ad absurdum* indicates that infinity comes in different sizes, and marked the definitive entry of infinity into the world of Mathematics.

II.4: Quantum/Classical Duality

Physics is a field where two different "ways of being" coexist, not without certain difficulties. At very small timescales and distances, physics (Quantum Mechanics) deals with phenomena (and an axiomatic body that explains them) radically different from the phenomena and axioms of "conventional" physics. Although quantum mechanics is beyond the scope of this text, it is important to note that the "quantum world" has some characteristics completely foreign to our normal experience:

- the ability of elementary particles to be both a wave and a particle
- quantum tunneling, which allows particles to pass through insurmountable barriers (for conventional physics)
- superposition phenomena, which allow the wave/particle to be in multiple states or places simultaneously
- quantum entanglement, which allows instant action at a distance

The theoretical model postulates an "ultimate basis" (wave equation, before observation) which, although it cannot be said to be analogic (it cannot be measured), is postulated to have a continuous probability distribution, formulated by the Schrödinger equation. However, as soon as we access it (observation and consequent wave collapse) reality becomes discrete ("Map"). Once in this Map-world, Thermodynamics (as statistical mechanics) acts as a "safety cushion," which not only guarantees the wave collapse (by the interaction between many particles), but is capable of statistically aggregating the behavior of many particles back into general, continuous, and analog laws, which constitute the backbone of Classical

Physics ("order from disorder," in Schrödinger's formulation (Schrödinger, 1944)).

The role of the observer remains contentious, as a necessary boundary for the leap between the quantum world and the conventional world (and this problem will be repeatedly addressed throughout this text). Moreover, although we currently live comfortably in a "classical" world, the entire theoretical body of Physics and Chemistry relies on quantum mechanics, and our lives increasingly depend on technologies directly based on it (transistors, lasers, magnetic resonance imaging).

Until now, Biology had remained isolated from this frontier: it was assumed that all biological phenomena operate in a "classical world." But even these barriers are falling (Al-Khalili & McFadden, 2014), and it seems that many crucial life processes (photosynthesis, respiration, even the very emergence of life) rely on quantum mechanisms without which life would be unviable.

II.5: Causality: A Map of the World as a Machine

Isaac Newton (1643–1727)

Arguably, one of the best paths to intellectual happiness is being able to build a bridge between the Territory and the Map. If this is the case, Newton must have been immensely happy at some point in his life. It is difficult to exaggerate the power and extent of his "maps." On one hand, his discovery (or invention?) of calculus allowed for the analysis of dynamic processes, turning Time into a conventional variable and thus allowing for its "mathematization." On the other hand, his law of Universal Gravitation explained nearly everything: from the kinetics of small objects to the movement of celestial bodies. It was genuinely Universal.

The success of his theory was so overwhelming that it became evident, and the axioms on which it was based became "obvious." Some among them are noteworthy:

- The existence of an absolute Space and Time, which "are out there," independent of the observer.
- The existence of phenomena that can be objectively observed, "from the outside," without being involved.
- The possibility of dividing phenomena into independent parts, assuming that understanding all parts separately equates to understanding the whole.
- Ultimately, the World as a gigantic Machine that obeys inescapable physical laws, universal in time and space. Differences that might arise between the predictions of the theory and the experimental results would only be attributable to errors in measurements or calculations.

Problems

However, problems soon emerged. Initially, these could be ignored, but eventually led the model to a profound crisis.

Relativity

The definitive crisis of the Newtonian model began in 1905 with Einstein's Special Theory of Relativity. Einstein challenged the concept of "simultaneous events" because this required a "signal" from the phenomena to the observer. This signal cannot be instantaneous, and thus its "ordering in time" depends on the distance: two events can be simultaneous or have different timing, depending on the observer's position (Einstein & Infeld, 1938). This theory destroys the idea of absolute and independent Space and Time and forces the consideration of a joint Space-Time, which is highly counterintuitive. However, this new approach explained certain "anomalies" not justified by the Newtonian model (e.g., the orbital motion of Mercury), and made some predictions slightly different from the Newtonian ones, which were overwhelmingly confirmed. Since then, the Newtonian model has remained as a "reasonable/simplified approximation," useful for "human dimensions," but essentially incorrect.

Quantum Mechanics

The Newtonian model is strictly deterministic. For this model, what we call "chance" or "randomness" is merely the insufficiency of our measurement methods or calculations. However, this idea was shattered when Quantum Mechanics demonstrated the existence of deep, irreducible randomness (e.g., in the decay of unstable isotopes or in wave function collapse). Ultimately, the classical model also did not admit quantized space and time; however, Quantum Mechanics postulates that there are "minimum units" of space and time (Planck length (1.6×10^{-35} m) and Planck time (5.4×10^{-44} sec)) below which time or space do not exist (i.e. it is senseless to consider time or space of these magnitudes).

Three-Body Problem; Nonlinear Dynamics

The classical Newtonian mathematical formulation required only two bodies, with the mass concentrated at their centers of gravity. Attempting to generalize this approach to more than two bodies opened a real Pandora's box. The "three-body problem" (Wikipedia: Three-body problem) proved to be mathematically intractable by conventional means. Additionally, one of the characteristics of the Newtonian model is that it is linear (no exponents >1). This implied that the initial inaccuracies ("margin of error") remained bounded and continued to be proportional to the variables under analysis. However, any attempt to generalize the model required the introduction of nonlinear terms (with exponent >1), which caused the initial errors to rapidly escalate, quickly invalidating any prediction. An approach to this problem by Henri Poincaré in 1899 opened up an entirely new science: Nonlinear Dynamics ("Chaos Theory").

The presence of nonlinear terms is not just a minor problem of accuracy in measurements: it is a fundamental limitation with enormous conceptual and practical consequences in terms of sensitivity to initial conditions, and emergent phenomena.

Sensitivity to Initial Conditions

Arbitrarily small differences in the initial conditions of a system can quickly lead to radically different situations. This is the phenomenon known as the "butterfly effect" (Wikipedia. Butterfly effect): alterations as small as desired (e.g., the flapping, or not, of a butterfly's wings in Brazil) can give way to very different outcomes (the appearance, or not, of a tornado in Ohio). It is impossible to define with infinite precision all the initial conditions of a system, and therefore a complete prediction is impossible. Once again, the concept of infinity (infinite precision) is the boundary (barrier/bridge) between the Map and the Territory.

Emergent Phenomena

Nonlinear systems can "jump" from one situation to another completely different one based on undetectable differences in initial conditions. The consequence is that game-changing behaviors, impossible to predict, may occasionally appear.

This is evident in Natural Sciences' scaling environments. It is not possible to predict the behavior of molecules based on the physical properties of atoms. The molecular level involves a qualitative leap (from Physics to Chemistry), which does not contradict the laws of Physics, but which cannot be formulated from that theoretical body. Similarly, the behavior of molecules does not contradict, but does not allow predicting the functioning of living beings (Chemistry -> Biology). Biology does not contradict Psychology, but it fails as a theoretical framework to analyze it.

General Systems Theory

The classical Newtonian model (linear, deterministic, observer-independent) allowed for a reductionist approach to problems: any complex problem could be subdivided into parts. If each part was well understood, the overall problem was solved.

However, the presence of nonlinear terms and the sensitivity to initial conditions made it impossible to manage situations with feedback loops, emergent phenomena, or multicausal processes, which are inherent to many phenomena, especially in Biology. An attempt (still nascent) to approach these problems is the General Systems Theory (Drack & Wolkenhauer, 2011).

This approach analyzes how the different elements of a system interact with each other, emphasizing the phenomena due to that interaction, rather than those due to each element. In sum, it studies what makes a set of elements more than the sum of its parts. This perspective, seemingly trivial, nevertheless has profound consequences:

- its focus on relationships (rather than elements) directly conflicts with our mentality, constantly centered on objects.
- given that it is impossible to conceive of a rigorously isolated system, a complete prediction is out of the question. Only an approximation within a limited range of space and time is possible. This is not a "technical," potentially solvable problem: it is an essential feature of scientific work.

Determinism

One of the corollaries of the "World-as-Machine" conception is the idea, almost universally accepted, that we live in a deterministic world, in which all phenomena have a cause. We accept without problem that this cause may not be known, and we are used to living within a margin of uncertainty, but we assume that, ultimately, every phenomenon has a cause. We usually label this uncertainty as randomness, but this does not challenge our deterministic conception: we usually attribute this randomness to the insufficiency of our capacity to collect data or analyze them. We tend to think that we could reduce this uncertainty by improving our data or our analytic power. Essentially, randomness would be a deterministic model that we fail to analyze, but there are no essential impediments to eventually understanding it. For example, we accept without problem that the vast majority of random number generation instruments are typical examples of deterministic algorithms.

However, ever more cracks are appearing in this conception.

We saw how quantum mechanics introduces a true, inescapable essential randomness that cannot be justified. There have been attempts to find causal mechanisms to justify this (so counterintuitive) randomness by resorting to "hidden variables" (Bohm, 1980), but in general, it is accepted that quantum randomness is irreducible.

Nonlinear dynamics does not break the deterministic model (in fact, one of its characteristics is being strictly deterministic), but it completely changes the rules of the game. The great advantage of linear deterministic models is that they allow exhaustive knowledge (and consequently, prediction) of the system's behavior. By introducing a nonlinear term, the behavior remains deterministic, but prediction quickly becomes impossible. Therefore, even though it is deterministic, it is actually unpredictable.

But the most spectacular cognitive blind spot of the conventional deterministic approach is the problem of free will. We have had no problem accepting that human behavior is (or should be) free, conditioned only by will (and this subjected to Reason, Ethics, or Faith, depending on cultures). All this, without the slightest hint of theoretical justification, and with plenty of experimental data against it (e.g., it can be demonstrated that an external observer can predict a "freely" made decision before the subject is aware of it (Libet, 1985)).

This probably poses a divorce between the theoretical and practical approaches. If I hesitate in choosing a dessert and finally decide to order ice cream, it is possible that my decision is conditioned by neuropsychological aspects (which "unconsciously" compel me), but the most effective description is that I just feel like having ice cream. This description will be the most effective and concise, although it is probably a huge oversimplification. At its core, it is the same problem faced by conventional physics ("quick and simple") versus relativistic physics (more complex, unnecessary at many scales, but "truer"). Some authors (Buonomano, 2017) suggest that perhaps free will is nothing more than a sensation obtained when making a decision (conditioned by factors which in many cases are unconscious).

In any case, the deterministic causality characteristic of a "World as a machine" model, typical of the Newtonian model, no longer holds, and it is significant that in all its cracks lies the problem of the observer. The deterministic causal paradigm is only applicable in the Map, and more specifically in the Newtonian Map (one could almost say it is the map of that Map).

And yet, it is difficult to escape from it.

Current Science, more than an attempt to explain the "why" of things, is an attempt to describe them as precisely as possible, and, above all, to make predictions about them (Bronowski, 1978). This optimizes the probabilities of survival and therefore carries a great selective pressure. This probably explains the enormous "causal craving" that we humans have: the irresistible tendency to seek a cause to explain any phenomenon (and allow predicting the consequences). Sometimes this generates paradoxes. For example, the existence of "retrograde causality" ("I have to leave now because my plane departs in two hours").

The problem of the observer becomes especially intractable in what has been called the *hard problem* of Neuroscience: the problem of consciousness (see The Problem of Consciousness, in I.5: "Consciousness as a Map"). Essentially, it would be about explaining *qualia*: phenomenological experiences. To grasp the idea of qualia, imagine a neuroscientist who knows everything about the physiology of color vision (Jackson, 1986). However, if this person is color-blind: she could know everything about red but not know red. This "redness" of red, the subjective experience (at its core, just the borderline between Territory and Map), is what the study of consciousness tries to address.

The Newtonian model of the World-Machine has allowed us to do "third-person" science for centuries, comfortably observing phenomena "from the outside." But this "third-person" approach cannot be transferred to the problem of consciousness: the observer is precisely what is being observed. Perhaps conventional science has not yet overcome this "objective" approach, and that is why it finds the problem of consciousness especially unapproachable.

II.6: Fractals

Why Fractals?

A fractal is a geometric object whose basic structure, fragmented or apparently irregular, repeats itself at different scales. Fractal-like structures are ubiquitous in nature (coasts, rivers, clouds, plants...), although obviously this pattern repetition at different scales operates within size limits. This is fundamentally different from mathematical fractals, which operate from 0 -> ∞.

There are several reasons to explore fractals in relation to the Territory/Map polarity.

1. They are an excellent example of a "platonic object," which exists in the Map, but of which we only have a "shadow" in the Territory.
2. In general, we tend to conceive Maps as a selective extraction of the Territory. The Map would in some way be an "essentialization" of Reality. However, there are situations where the Map-world exceeds the Territory-world. A paradigmatic example is fractals. As discussed in II.3: "Differential Calculus" or II.7: "Fourier Transform," the leap from Territory to Map is often made through the "conceptual bridge" of infinity. Fractals (like many other mathematical objects) are another example. But, unlike conceptualization mechanisms or the Fourier transform, in the case of fractals the Map exceeds the Territory, and the latter is just a rough approximation to the "platonic" original.
3. Fractal geometry has properties that reflect the characteristics of the boundary between M and T: they are two clearly distinct worlds, but inextricably intertwined. They cannot be mixed, but it is impossible to draw the line of demarcation.

4. In Medicine, they are an example of the failure of an obsolete Map (III.5: "Fractal Pathology"), and of how we need to create a new Map to navigate new Territories.

History

In the late 19th century, atypical mathematical objects (Cantor set, Julia set, Koch curve) began to be described. They posed some uncomfortable problems for traditional mathematics but were treated as "anomalies" in the sense formulated by Thomas Kuhn (Kuhn, 1996).

In general, they are the result of recursive processes. For example, for the Koch snowflake, the process is very simple:

1. Take a triangle with side n
2. Divide each side into 3 equal parts
3. Replace the middle section with the two edges of an equilateral triangle of length $n/3$
4. Repeat this process on each subsegment indefinitely

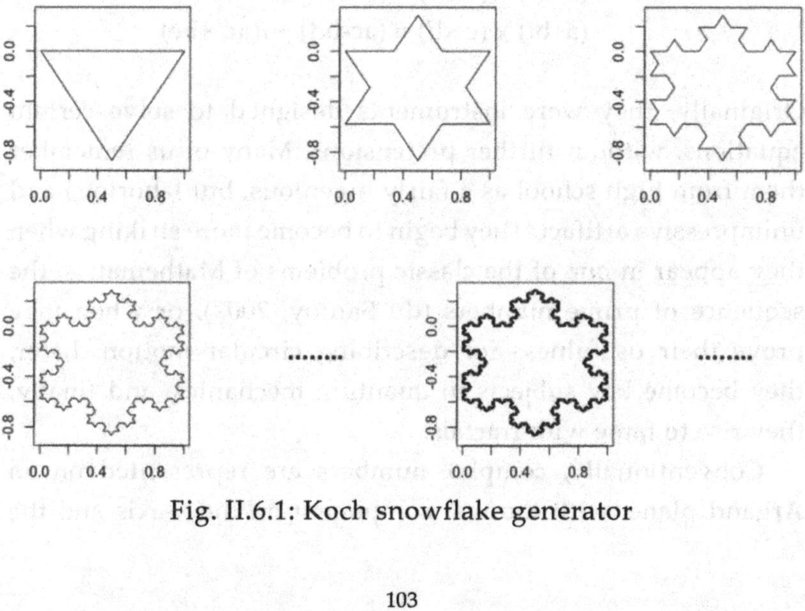

Fig. II.6.1: Koch snowflake generator

Fractals have curious mathematical properties: for example, the Koch snowflake has a finite area (in our figure, evidently $< n^2$), yet an infinite perimeter.

The Mandelbrot Set

The advent of computers, and with them the exponential increase in computational capacity, changed the scenario. It became very easy to program and explore very prolonged recursive algorithms.

A different thread that contributed to the emergence of the fractal concept was the mathematics of complex numbers, composed of a real part and an imaginary part:

$$z = a + bi$$

where $i = \sqrt{-1}$ (which is impossible in conventional arithmetic). The rules for operating complex numbers are similar to those of conventional algebra:

$$(a+bi) + (c+di) = (a+c) + i(b+d)$$
$$(a+bi) \times (c+di) = (ac-bd) + i(ad+bc)$$

Originally, they were instruments designed to solve certain equations, without further pretensions. Many of us remember them from high school as a fairly ingenious, but laborious and unimpressive artifact. They begin to become more striking when they appear in one of the classic problems of Mathematics: the sequence of prime numbers (du Sautoy, 2007), or when they prove their usefulness for describing circular motion. Later, they become key subjects in quantum mechanics, and finally, they rise to fame with fractals.

Conventionally, complex numbers are represented on an Argand plane, with the real component on the x-axis and the imaginary component on the y-axis. Thus, the number (4+2i) would be represented as:

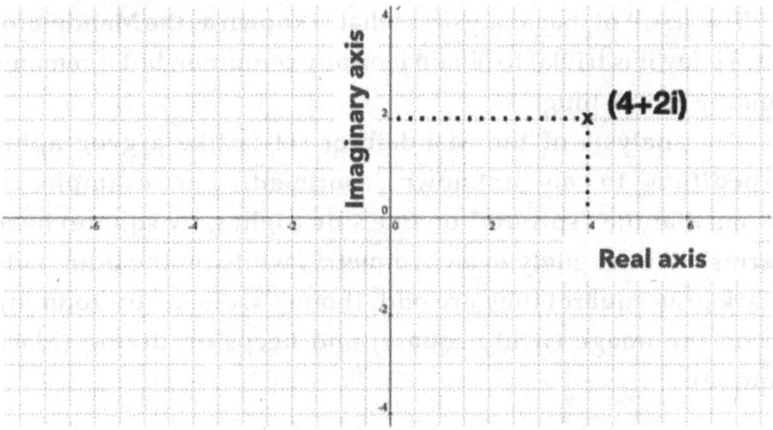

Fig. II.6.2: Argand plane: complex number (4,2)

In 1972, the mathematician Benoit Mandelbrot faced an apparently trivial problem. He analyzed a recursive application:

$$z_{i+1} = z_i^2 + c$$

where *z* and *c* are complex numbers. In summary, it is a series that starts at the origin (z = 0 + 0*i*) and in which each new term is obtained by squaring the previous one and adding a constant (following the rules of operation for complex numbers). He soon saw that, depending on the value of *c*, the resulting series in some cases would shoot off to ∞, while in other cases it remained bounded.

Mandelbrot attempted to create a map on the Argand plane, marking those values of "c" which caused the series to shoot off to infinity and those which kept it bounded. At that time, this required substantial computational capacity and a prolonged time, so after setting the process to run, he left the computer. When he returned, he found such extraordinary drawings on the screen that he assumed the computer or the screen had broken.

The result of that analysis is what is known as the Mandelbrot set. Today it is trivial to obtain it on any computer, but it remains equally astonishing.

The analysis of the Mandelbrot set is like a geographic expedition to an unknown continent. For example, it seems that the "sprouts" on one side (light gray square) have "arms" with vaguely even symmetry, while on the other side (dark gray square) they are odd. If one "dives in" by zooming in on the image (white square) and keeps on diving (black square):

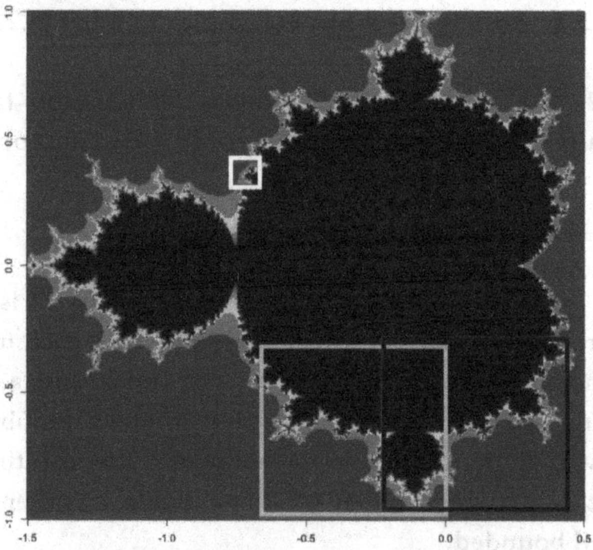

Fig. II.6.3: Mandelbrot Set: In black, the points that remain bounded; in gray, the points that shoot off (different tones of gray, depending on the rate of dispersion). (Figure obtained using a script in R.)

Fig. II.6.4: Mandelbrot set (detail 1)

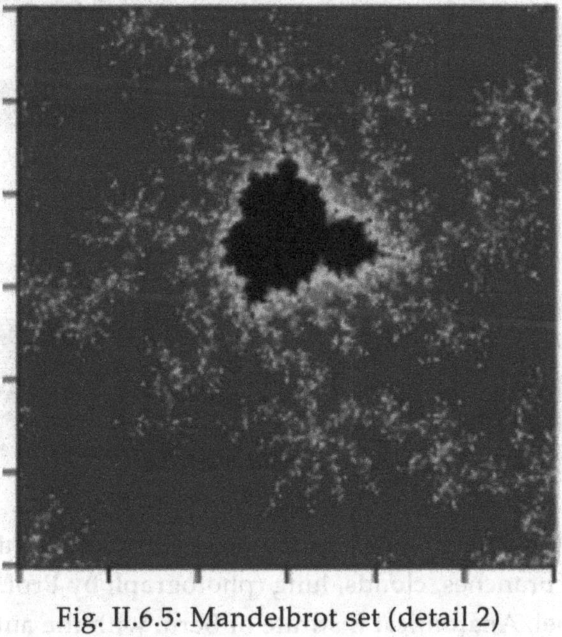

Fig. II.6.5: Mandelbrot set (detail 2)

A new shape appears that is similar (but not identical) to the original. This exploration can be continued infinitely. A good "dive" can be seen at: https://en.wikipedia.org/wiki/Fractal#/media/File.gif.

It is difficult to believe that this set was "invented" by anyone. It seems more intuitive to interpret it as a kind of unknown "platonic object" until its discovery by Mandelbrot.

This discovery opened up a whole new field in geometry. Mandelbrot called it a fractal and pointed out how this type of geometry is ubiquitous in all realms of nature (Mandelbrot, 1983).

Furthermore, using fractal geometry techniques it is easy to artificially produce images that are almost indistinguishable from nature. (In fact, this is how most science fiction "landscapes" are produced.)

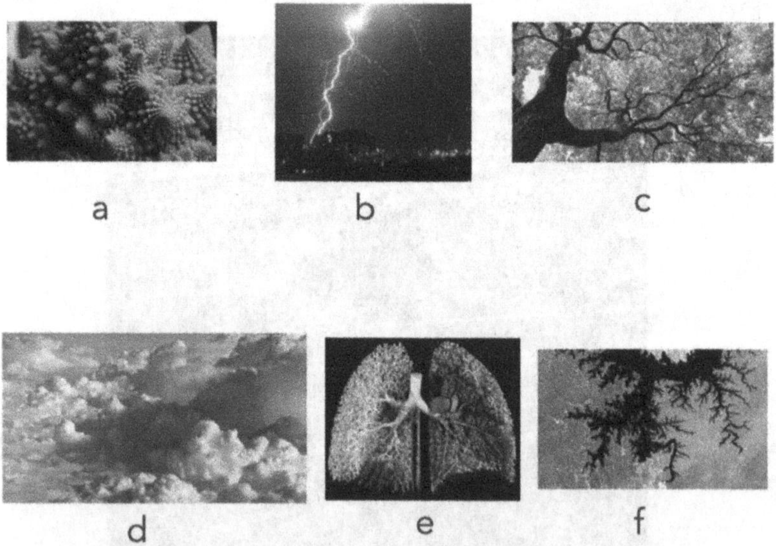

Fig. II.6.6: Fractal objects in nature: broccoli, lightning, tree branches, clouds, lung (photograph by Prof. E.R. Weibel, Anatomical Institute of Bern, with the author's permission), image of the Nile River in As Subu (Egypt). (Image obtained with Google Earth.)

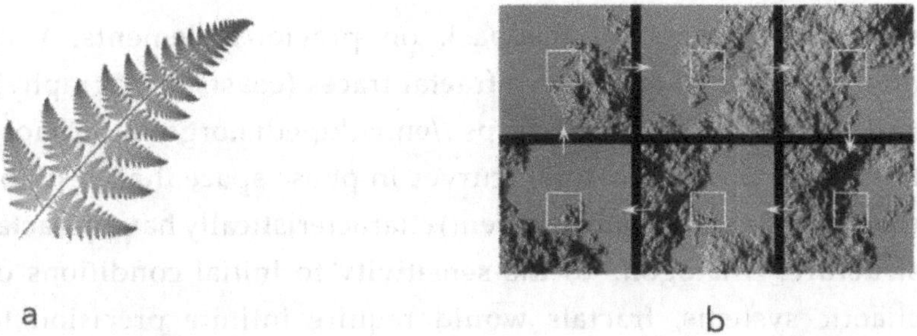

Fig. II.6.7: Fractal images produced by computer. Fern leaf created with an R script based on Barnsley's algorithm; recursive image of a fractal coast downloaded from Pászto, Vít, Marek, Lukas, & Tucek, Pavel. (2011): Fractal dimension calculation for CORINE land-cover evaluation in GIS – A case study. *CEUR Workshop Proceedings*. 706. 196–205, with the authors' permission.

Fractal geometry has some characteristics that explain its widespread presence in nature.

1. It is a good "dimensional exchanger." When a three-dimensional structure (e.g., a transport tube: 3D) becomes a two-dimensional structure (e.g., a diffusion or exchange surface: 2D), the interface is almost always a fractal. This is true in the bronchial system, the capillary system, all portal systems, renal glomeruli, etc.
2. It is a very compact encoding and transmission system. Very few lines of code (computer or genetic) can produce very complex structures.
3. It is a very robust system. Damage to one part only affects the "descendants," but does not generalize.
4. It is very flexible and easy to adapt to different volumes or spatial forms.

Fractals have an intimate relationship with nonlinear dynamic systems ("deterministic chaos"). Both are the product of recursive systems, of feedback on previous elements. Very often, chaotic systems leave fractal traces (coasts, topography), and strange attractors (https://en.wikipedia.org/wiki/Chaos_theory#Strange_attractors) (curves in phase space that describe the trajectory of a chaotic system) characteristically have a fractal structure. Analogous to the sensitivity to initial conditions of chaotic systems, fractals would require infinite precision to catalog certain points (e.g., in the vicinity of (-0.5 -0.5i) of the Mandelbrot set (Fig. II.6.3)). And they are not figures "that are there": the mathematical fractal is an infinite recursive process: we can never have a "complete fractal"; natural fractals are only approximations. As with biological life, it does not "exist": it "happens."

Fractals in Geography: Fractal Dimension

Probably what brought Mandelbrot to fame was a very short article in *Science* in 1967 (Mandelbrot, 1967), which planted the question, "How long is the coast of Britain?" This was a topic that had already been raised before: in 1961, Lewis F. Richardson (Richardson, 1961) had commented on how the border between Spain and Portugal measured 987 km, or 1214 km, depending on "which side was measured" (which country was considered). Similarly, the border between Holland and Belgium measured 380 km or 449 km, depending on the side. Mandelbrot's conclusion was that, obviously, it depended on how it was measured (Fig. II.6.8).

The basis of Mandelbrot's observation was that there was an inverse exponential relationship between the length of the "measuring stick" and the result. The smaller the measuring stick, the (exponentially) greater the resulting perimeter. Think about how different the results would be obtained by tracing the coast on the map with a pair of compass dividers (and it would depend on the scale of the map and the size of the compass step), or walking it on the ground, or meticulously tracing every small stone…

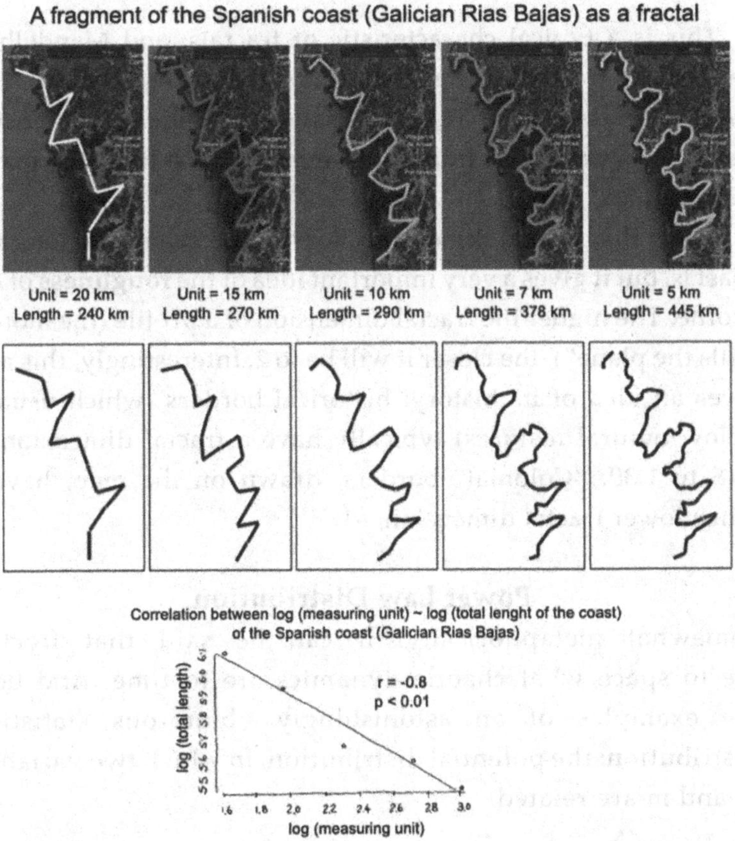

Fig. II.6.8: Fractal Dimension of the Coast of the Galician Rías Bajas, by the Compass Method. The total length of the coast is analyzed according to the size of the "compass step" used. If the structure has fractal characteristics, there will be a linear correlation between the logarithm of the unit of measure and the logarithm of the total length obtained (in this case, $r = 0.98$, $p < 0.01$). The fractal dimension of the coast is obtained from the slope of the regression line.

Since the relationship between the step size and the result obtained is (inversely) exponential, the relationship between the logarithm of the unit of measure and the logarithm of the result is linear (Fig. II.6.8).

This is a typical characteristic of fractals, and Mandelbrot defined the fractal dimension of this type of structure based on the slope of that curve. This breaks the classic idea of dimension necessarily expressed in natural numbers (1D: line, 2D: plane, 3D: volume).

Naturally, fractal dimension does not answer how long the coast is, but it gives a very important idea of the **roughness** of the profile. The higher the fractal dimension of a profile (the more it "fills the plane"), the closer it will be to 2. Interestingly, this also gives an idea of its history: historical borders (which usually follow natural features) typically have a fractal dimension of 1.15 to 1.30. "Colonial" borders, drawn on the map, have a much lower fractal dimension, ≈1.

Power Law Distribution

Somewhat metaphorically, it can be said that fractals are to space what chaotic dynamics are to time. And both are examples of an astonishingly ubiquitous statistical distribution: the potential distribution, in which two variables M and m are related:

$$M \sim m^p \quad \text{(and thus: } \log(M) \sim p \log(m))$$

where p is the slope of the log ~ log regression line.

The ubiquity of this relationship is fascinating. To cite just a few examples:

- The Pareto distribution (relationship between each person's wealth and the number of people with that capital) (Fig. II.6.9.A)
- Zipf's law (relationship between the frequency of a word's appearance and the number of words with that frequency) (Fig. II.6.9.B)

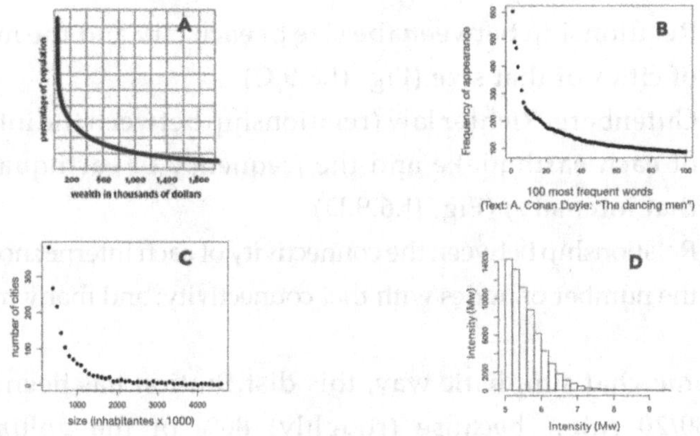

Fig. II.6.9: Examples of Power Law Distributions. A: Pareto Distribution. Economic resources per household in the USA in 1997. Source: Buchanan, M. "Wealth happens." *Harvard Business Review*, April 2002 (with permission from the publisher). B: Zipf's Law. Frequency of appearance of each word in a text. Text: A. Conan Doyle: "The Adventure of the Dancing Men." Analysis conducted with ChatGPT. The 10 most frequent words were: "the," "and," "to," "of," "a," "I," "in," "that," "was," "it." C: Distribution of city sizes. On the horizontal axis, city size (in number of inhabitants, grouped in blocks of 100,000 inhabitants). For example, there are 380 cities with between 300,000 and 400,000 inhabitants, 31 cities with between 3.1 and 1.4 million inhabitants, 21 cities with between 2.3 and 2.4 million inhabitants. There are only 3 cities with more than 25 million inhabitants (Tokyo, Delhi, and Shanghai). Data obtained from the United Nations Department of Economic and Social Affairs (https://population.un.org/wup/Download/), processed with R. D: Gutenberg-Richter Law. Frequency distribution of earthquakes with intensity >5 on the Richter scale worldwide from 1904 to 2019. Horizontal axis: earthquake intensity (in Moment Magnitude); vertical axis: number of earthquakes of that intensity. For example, 53,337 earthquakes were recorded with Mw between 5 and 6, 1185 with Mw between 7 and 8, and only 4 with Mw >9 (Chile 1960, Alaska 1964, Sumatra 2004, Tohoku 2011). Data obtained from the International Seismological Centre (http://www.isc.ac.uk/iscgem/download.php), analyzed with R.

- Relationship between the size of each city and the number of cities of that size (Fig. II.6.9.C)
- Gutenberg-Richter law (relationship between the intensity of each earthquake and the frequency of earthquakes of that intensity) (Fig. II.6.9.D)
- Relationship between the connectivity of each Internet node and the number of nodes with that connectivity; and many more.

In a somewhat simplistic way, this distribution has been called the "80/20 rule," because (roughly) 80% of the volume (of wealth, or words, or connections, etc.) is in the hands of 20% of the sample (of people, of terms, of nodes, etc.).

In general, this type of distribution indicates that the analyzed structure or dynamic is complex, with multiple recursive circuits (and, consequently, whenever one encounters this distribution, it should be assumed that *something interesting is happening*).

The exponent of the curve (or, in other words, the slope of the line that plots the logarithms of the two variables) provides valuable information about the studied dynamic (e.g., about the distribution of wealth in a country, or about seismic risk in an area, or about the richness of vocabulary, etc.).

II.7: Fourier Transform

A paradigmatic example of the leap from Territory to Map is the Fourier transform. An excellent explanation of the process can be found at: https://www.youtube.com/watch?v=spUNpyF58BY

In summary, it consists of taking a time series (a variable that fluctuates over time), "collapsing" time, and thus omitting the transient aspect: having all time visible at once. In short, "freeing it from territoriality" and converting it into a transportable, modifiable, and graftable object in other contexts. In some way, it is the process of making a Map of the time series.

We will first do a very superficial theoretical review, using sound as an example, and then analyze the consequences of the process.

Recording and Recovering Sound

Sound is pressure waves transmitted through the air. When they reach the eardrum, they are collected by a membrane that processes them and converts them into electrical signals. We will not discuss the neurophysiological aspect and will remain at the internal ear level.

The first way to record and recover these waves was to convert them into irregularities on a smooth surface. When these irregularities were "read" with a needle, the pressure waves in the air were reproduced, which could be recognized as sounds similar to the original ones. This was the basis of the first slate or Bakelite records.

Fourier Transform

When a sound reaches the eardrum, what actually arrives are a series of rises and falls in air pressure (Fig. II.7.1).

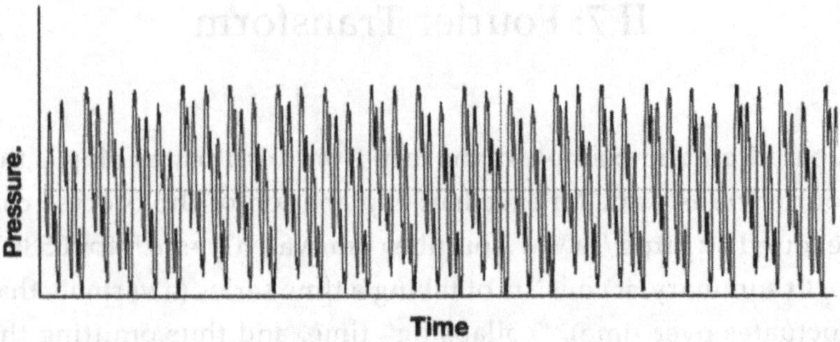

Fig. II.7.1: Pressure wave. X-axis: time. Y-axis: pressure. (Image obtained from ambient noise with Praat.6.0.30 (www.praat.org).)

In summary, it is a time series of pressure measurements.

Fourier demonstrated that any time series could be represented as the sum of a series of sinusoidal waves.

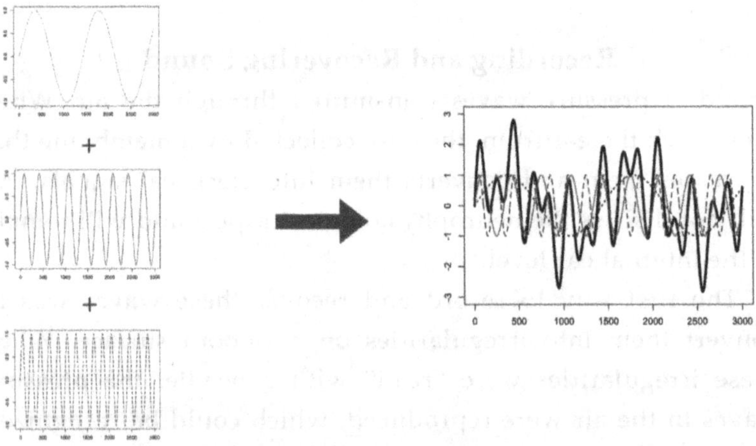

Fig. II.7.2: Complex wave, sum of three sinusoidal series (freqs 0.03, 0.05, 0.007 Hz-, in different gray tones, and sum of the previous three series (black, thick line)). (Image obtained with an R script.)

The Fourier transform is a mathematical operation that allows the extraction of the sinusoidal series that form the complex wave. It measures how *strong* the contribution of each frequency is, as well as its position in time relative to the other frequencies. In summary, it converts a complex time series into a set of frequencies. For example, the phrase "one, two, three," represented as a sound wave, would be:

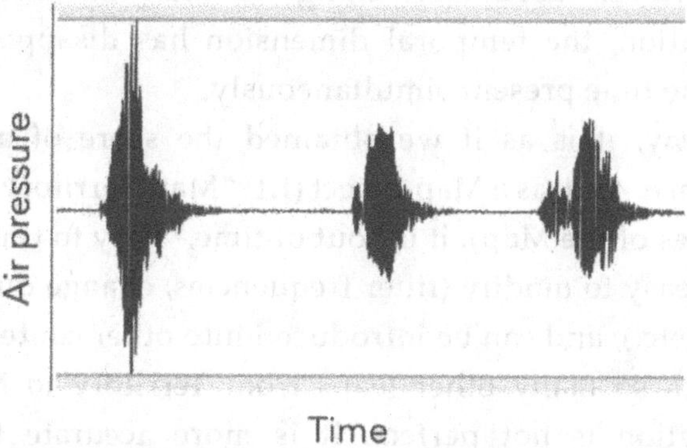

Fig. II.7.3: Time series in real-time mode: recording of the phrase "one, two, three." X-axis: time; Y-axis: pressure. (Image obtained using Praat.6.4.13 (www.praat.org).)

After the Fourier transform, the result is:

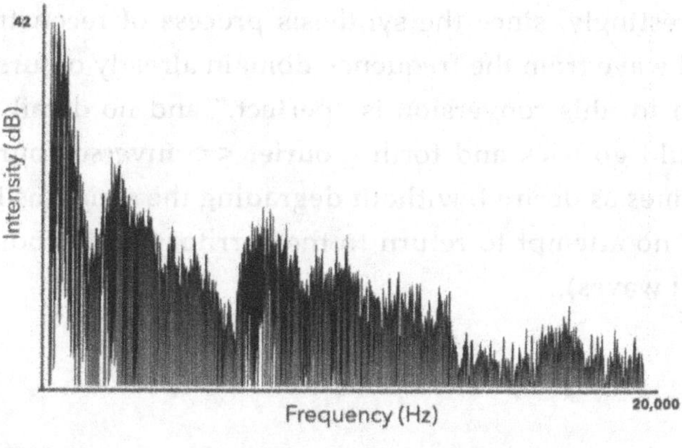

Fig. II.7.4: Frequency mode of the sentence "one, two, three." X-axis: frequency; Y-axis: intensity. (Image obtained using Praat.6.0.30 (www.praat.org).)

In the second graph, the x-axis is the frequency (in Hz), and the y-axis is the intensity (in dB). The first graph is the representation in the time domain, and the second in the frequency domain. Using this information, an inverse process (synthesis) can reconstruct the original sound, as if by magic.

But the extraordinary thing is that in the second representation, the temporal dimension has disappeared: we have all the time present simultaneously.

In a way, it is as if we obtained the score of a musical performance. And, as a Map-object (I.1: "Map/Territory," section Advantages of the Map), it is "out of time," easy to transport or transmit, easy to modify (filter frequencies, change duration or intensity, etc.), and can be introduced into other contexts.

As with so many other leaps from Territory to Map, this representation is not perfect. It is more accurate the more exhaustive the number of frequencies recorded, but there will always be a slight difference. A perfect representation of any real-world sound would require sampling infinite frequencies. As in so many other leaps between Map and Territory, the concept of infinity is the bridge/barrier that connects/separates the two worlds.

Interestingly, since the synthesis process of reconstructing a sound wave from the frequency domain already occurs in the Map-world, this conversion is "perfect," and no detail is lost. One could go back and forth (Fourier <=> inverse Fourier) as many times as desired, without degrading the series (as long as there is no attempt to return to the Territory and produce the physical waves).

II.8: Renormalization and Complexity

A discipline that specifically explores the leap between Map and Territory is Renormalization studies (DeDeo, n.d.). Essentially, these techniques explore the processes of simplification (coarse-graining) and their consequences. The idea of coarse-graining (strictly speaking, increasing the "grain" of an image) is to reduce the amount of information carried by a message as much as possible without losing meaning. In short, it is a system of "pixelation."

A classic example would be the process of digitizing an image: how to "compress" an image into a Map-object made up of a series of ones and zeros so that it can be filed, transmitted, and manipulated. Another typical example would be Economic Theory: how to extract rules from the set of economic exchanges of a system that allow the description and analysis of its operation. The same problem arises in certain areas of physics, neuroscience, and in general when attempting to abstract a complex model.

And, naturally, problems occur on the way back: from some data, we have obtained a "map" that allows us to describe the entire system. But how well does it fit? To what extent do the model's predictions hold? How far is the real situation from the model after a certain time?

Renormalization and complexity analysis examine both processes: how it is pixelated, and what is lost along the way. Again, the concept of entropy appears as the distance between the Map and the Territory: as what is omitted by modeling.

Complexity in Biomedicine

In general, it is considered that a healthy biological system is an orderly and predictable system. In fact, in many languages, the term "disorder" is used as a synonym for disease. However, the reality is the opposite: healthy systems are characterized by being "disordered" and unpredictable. The appearance of rigid and predictable behavior (e.g., the appearance of rhythms) is often a sign of disease (Parkinsonian tremor, cyclothymia, epilepsy, Cheyne-Stokes periodic breathing, etc.). In many biological systems, one of the earliest signs of disease or aging is the loss of complexity in the system's functioning (Lipsitz & Goldberger, 1992; Vargas et al., 2015). Applied to time series (heart rate, temperature, blood glucose, oximetry, etc.), it is becoming increasingly useful clinically to be able to objectively measure the complexity of the recording as an indicator of the system's health. Although there are many techniques to assess this variable, most are based on two approaches:

- **Measures Derived from Entropy** (Approximate Entropy, Sample Entropy, etc.)
  Essentially, they evaluate to what extent the knowledge of a certain number of successive points of the time series allows us to predict what the next point will be — in short, to what extent the evolution is "predictable." A healthy system will be able to adapt quickly to changes, and therefore its trace will be very "rough" and difficult to predict precisely: its fit to our "map" will be poor, and predictions will be imprecise. Conversely, a diseased system takes longer to react and presents a simpler and more predictable trace, and is associated with a poorer prognosis (Varela et al.,2006).
- **Information Distribution Measures** (Detrended Fluctuation Analysis)
  Another way to tackle the problem is to analyze how information is distributed across different window sizes. A structure will be more complex the more it depends on the "fine" details perceived in small windows. This can be seen by comparing the information lost from a

photograph taken from 1 km or from 30 km altitude in a complex structure — e.g., the city of Venice (Fig. II.8.1.A) versus photographs taken from the same distance in the Sahara Desert (Fig. II.8.1.B).

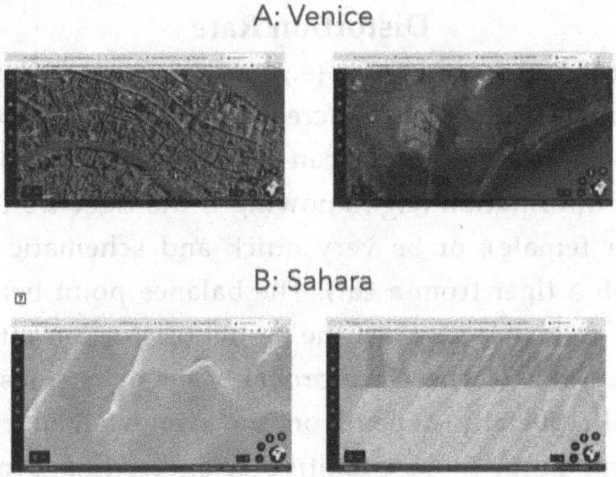

Fig. II.8.1: A: Photograph of Venice from an altitude of 1 km (left) or 30 km (right). B: Photograph of an area of the Sahara from an altitude of 1 km (left) or 30 km (right). (Images obtained from Google Earth, https://www.google.es/intl/es/earth/index.html.)

Similarly, a time series will be more complex the more information is encapsulated in short time windows, or in other words, the less time it takes for the influence of each point to "dissipate." In a river with a smooth and calm flow (low complexity), the turbulences induced by a small obstacle can be seen for many meters. In a fast and turbulent river (high complexity), the turbulences induced by an obstacle are quickly lost. There are mathematical techniques (e.g., DFA, detrended fluctuation analysis) to analyze how long the influence of each point lasts in a time series, and overall it has been seen that this measure of complexity is also an early marker of pathology or senility (Churruca et al., 2008; Lundelin et al.,

2010; Rodríguez de Castro et al., 2017; Varela et al., 2015). (A summary of DFA can be found at https://drive.google.com/file/d/1fqWLJ_2iBvPk7IAGssiefNArfxo-EvdG/view?usp=drive_link.)

Distortion Rate

When making any type of map (e.g., a situational judgment), the increase in accuracy has an increasingly higher computational cost. A very precise model can be very costly and provide irrelevant information (e.g., knowing if the tiger we encounter is male or female), or be very quick and schematic (and not distinguish a tiger from a cat). The balance point between the efficiency and complexity of the model (distortion rate theory) is a critical choice. There are probably no strict rules, and the choice will crucially depend on the context. Naturally, this is a central problem in cognitive sciences, image or speech recognition, etc., and will be a central theme in all the leaps between Territory and Map.

II.9: The Map/Territory Polarity in Ethics

The Map/Territory polarity is also present in Ethics.

There exists a Map-Ethics, principlist, based on axioms that underpin norms of conduct and are not questioned. For example, in medical Bioethics, four axiomatic principles are postulated: non-maleficence (the patient should not be harmed), beneficence (the patient should benefit), autonomy (the patient has the right to choose, considering his/her values), and equity (consideration should be given not only to the patient but to the entire population). Ethical conduct would be the optimization and harmonization of these principles.

On the other hand, there is a Territory-Ethics, consequentialist, which posits that conduct should be guided by the expected consequences of one's actions.

Often, this does not imply a contradiction (e.g., if one seeks to optimize the aforementioned principles). But this is not always the case, and it is not uncommon for there to be contradictions between these principles, typically between the principle of beneficence and that of equity: the archetypal examples being organ transplants or the public management of very expensive treatments. In these cases, the principle of beneficence (the physician should be guided by the best interest of the patient) conflicts with the limitations of a scarce or very expensive resource, which must be distributed in society following criteria of equity (the whole of society must be taken into consideration). This is a thorny problem for the principlist approach.

The classic example of collision between principlist and consequentialist ethics is the "trolley problem," where an observer has to choose between behaving actively (diverting a trolley and consequently killing one person) or doing nothing (and consequently letting several people be run over and killed).

Fig. II.9.1: The Trolley Problem

This dilemma between the principlist and consequentialist approaches is present in many of the ethical conflicts affecting clinical practice. Antiabortionists defend the principle of the right to life. Proponents of legalization point to the risks to the mother and the dangers of illegal abortions.

This same Map-Ethics versus Territory-Ethics dilemma is seen in numerous political conflicts. In Spain, those who reject Catalonia's independence do so based on the constitutional principle, while those who advocate independence point to the impossibility of governing against a citizen majority.

The same dilemma affects technological changes. In self-driving cars, it pits cars that will prioritize minimizing overall casualties (principlist) against those that will prioritize saving the owner (consequentialist, and likely chosen by buyers).

In summary, it would be an example of the dialectic between the Map-world (rules and reasons) and the Territory-world (causes and consequences).

II.10: The Poles

The fundamental thesis of this text is that there are two "modes" of being, which schematically could be called "Map" and "Territory" (I.1: "Map/Territory"). They are not easily definable (because when trying to define them, we automatically switch to "Map mode"), but they have some differential characteristics (Table II.10.1) and it is not difficult to find typical examples of each mode.

It is also easy to find examples of the leap from Territory -> Map (e.g., classification, conceptualization) or, conversely, Map -> Territory (e.g., building from a blueprint, playing a musical score).

Table II.10.1

Territory mode	**Map mode**
Continuous	Discrete
Analogic	Digital
Context-dependent logic	Excluded middle logic
Heuristic learning	Algorithmic learning
Causes	Reasons
.....	.....
Examples	
Musical interpretation	Musical score
Cooked dish	Recipe
Theatre performance	Text
.....	.....

A key idea in Taoism is the immiscibility and inseparability of the two poles, yin and yang. The same is true for the Map and Territory modes. We will see that it is impossible to find an example of a pure Map or pure Territory. But first, let's explore some polar examples of each mode.

1. Anchoring in the Territory

The brain has an irresistible drive to make Maps, and it is impossible to find a pure Territory area. But there are some situations or activities where we are anchored in the Territory, and the Map tends to blur.

Meditation

Probably the term "meditation" is unfortunate. Certainly, one does not "meditate" on anything. Although there is no strict definition of what meditation is (Emmanuel Carrère offers 26 different definitions, Carrére, 2021), surely most would agree that it is the voluntary attitude of trying to stay "in the present moment," without memories, without plans, without expectations. Generally, it is suggested that the meditator fix his/her attention on something (breathing, a sound, a flame) and concentrate exclusively on that aspect and the sensations that arise. In summary, an attempt to stick strictly to the Territory. Naturally, this is not a situation; it is a (long) process, in which the meditator tries to "turn off" distractions (*vrtti*, fluctuations of the mind): what in meditators' jargon is called "monkey mind," a mind that jumps from topic to topic, almost always with self-referential features.

Flow

"Flow" ("the zone") is a psychological state described by artists, elite athletes, or creators, in which the subject is intensely involved in a technically difficult activity but executes it perfectly, with a great sense of ease and joy, with minimal self-consciousness, and frequently with an altered sense of time passing. A classic example would be a musician improvising in full inspiration. One could say that the subject is immersed in the Territory, without plans, without worrying about the past or the future, perfectly focused on his/her activity.

Sex

Without any normative intent and accepting as much variability as one wishes, it could be said that in sexual relations (and especially in orgasm), there is a moment of relaxation of self-consciousness and temporal perception: somehow, a pure Territory.

Of course, none of the above are pure "Territory situations." It could be argued that eroticism or pornography is the Map component that slips into sexual activity, and one of the characteristics of flow is the naturalness and ease with which the performer adjusts to the "map" (sports rules, music characteristics, etc.), but they are situations where the conscious aspect of the map tends to blur.

Several common characteristics are found in these three states:

- Intense concentration on the activity performed, without paying attention to other stimuli, and with a sense of ease and pleasure. This corresponds to the Taoist idea of *wu wei*: passive, effortless action.
- An autotelic feeling: the activity is its own motivation; it does not seek rewards or external stimuli.
- A decrease in self-referential thoughts, with a blurring of individuality, which tends to merge with the activity performed.
- An alteration in the perception of time.
- A decrease in verbal-analytical control of motor skills: one "does without having to think."

Most of the literature on these topics is written in subjective, "first-person" terms, and it is difficult to subject it to conventional scientific analysis. Our scientific mindset is anchored in the Newtonian paradigm of the "World-Machine," which is analyzed and studied "from the outside," and we have few instruments to analyze the "object + observing subject" set (see II.5: "Causality: A Map of the World as a Machine"). However, neuroscience is beginning to have some tools to study these phenomena "from the outside." And a common feature that robustly repeats in all these situations of anchoring to the Territory is a depression of the activity of the Default Mode Network (DMN) (Brewer et al., 2011; Georgiadis & Kringelbach, 2012; Harris et al., 2017; van der Linden et al., 2021), which can be measured by fMRI (functional magnetic resonance imaging).

The DMN was introduced in I.3: "Mind/Map versus Brain/Territory." It is a network of brain areas with functions of general coordination of the mental process, specifically related to self-referential thoughts, manipulation of the temporal dimension (memories, projects), theory of mind ("what will others think?"), and the sense of individuality. Some authors (Carhart-Harris & Friston, 2010) identify it as the structure responsible for the Freudian ego, and it is not surprising that meditation, flow, and sex are some of the classic mechanisms in which ego dissolution is described. The DMN is also responsible for the mind-wandering or monkey mind that meditation tries to evade. The parallel between these situations and psychedelic and mystical experiences is striking, both in phenomenological aspects and in DMN inhibition.

2. Sticking to the Map

The opposite pole would be to seek a pure Map: a kind of uncontaminated meme.

Cartography

It would seem that an archetypal example of "Map mode" would be maps themselves: a "discretized" object, where what is included and what is not has been chosen, and which would seem completely objective, independent of context.

John B. Harley warns about how fallacious this view is (Harley, 2001). Maps are promoted by patrons and made from a certain perspective, which selects how they are centered (preferably on the northern hemisphere, western European zone), how they are represented (generally Mercator projection, which tends to hyper-represent some areas and relatively diminish others), what appears (border lines, cities, roads), and what is omitted (military installations, "unimportant" structures). An extreme case is gerrymandering, planning territorial organization to optimize electoral results for certain political parties.

There is no objective map from the moment a selective extraction (by what criteria?) of the information from the Territory is decided. Historically, maps have been instruments at the service of power, decisively conditioned by the concepts of the Nation-State, colonial mentality, and land as property.

Additionally, the Map becomes a Territory-object in itself (and can be stolen by a spy or sold as a valuable antique).

Chess Game

The game of chess could also seem like a Map-object: well-defined rules, no room for chance (aside from who opens the game). In short, a decontextualized space, pure algorithmic thinking.

Yet, it is not so. The game is inevitably loaded with affective aspects (aggressive versus defensive players, struggle to maintain the initiative, strategies to push the time limit, etc.).

It is doubtful that there exists (for now) an unbeatable pure algorithmic instrument. In 1997, Deep Blue defeated world champion Garry Kasparov, but, contrary to what it might seem, this was not the definitive victory of the algorithmic approach. Although obviously, Deep Blue's "hard" analytical capacity was incomparably superior to any human's (it could analyze between 100 and 300 million positions per second), it needed non-algorithmic elements to narrow down searches (and economize computing power). These elements are "Territory variables," not obtained by the system, to which heuristic "weights" are assigned (e.g., control of the center, pawn structure, degree of king's cover, etc.) (ProfessionalAI.com, 2020).

In III.6: "Sex, Race, Species, Gene," other examples of idea-maps that unravel when analyzed closely are discussed.

II.11: Et cetera…

Politics is another field where the Map and Territory polarity is displayed.

A naive first approach to Democracy (and to bourgeois revolutions in general) is the idea of government by the People, as opposed to a pre-established and unquestionable authority (King, Pope, etc.): in a way, rooting power in the Territory.

The very basic political concepts have their polar tint: although there is overlap, the idea of Nation is more Territory (in the sense of complexity, contextuality), while State is more Map, more concept[1] of organization and structure.

This polarity is clear in the revolutionary field. The First International split in 1872 was due to the clash between Bakunin's anarchist, self-managed, and socially transformative conception (Territory), and Marx's centralist, political, efficiency-optimizing idea (Map). The conflict reappeared between anarchists and communists in the Spanish Civil War and is present in the struggles between defenders of the public versus the common, or direct democracy versus representative democracy.

Jean-Claude Milner's description of political representation (Milner, 2013) also reflects this polarity. Milner establishes a parallel between politics (as practiced in the "Western democratic world") and Greek tragedy. There would thus be an audience that identifies (mimesis) with politicians/actors, but always maintaining the barrier (the "fourth wall" of the theater). The model would work as long as there is empathy between the actors and the audience (but without the audience trying to get on stage).

A clear example of the changes that crossing that fourth wall implies could be the evolution of the Spanish 15M movement to Podemos. The 15M was a mass, self-managed movement, "Territory" (one of its slogans was: "Occupy the squares!").

When, in the interest of greater efficiency to produce changes, a significant part of this movement jumped to institutional politics (formation of Podemos as a political party), many of the characteristic changes of the leap from Territory to Map occurred: by obtaining the benefits of the Map mode (efficiency, simplification, ability to generalize, communicate, and extrapolate, etc.), a price is paid in the richness of nuances, the movement is less participatory, and the overall discourse shifts from being bottom-up to top-down. Possibly, a similar phenomenon can be seen in other political movements of the early 21st century (e.g., Syriza in Greece, Arab Spring in northern Africa, Gilets jaunes in France, etc.).

This politics-theater model has consequences that we may be seeing today. On one hand, it only works whilst there is this empathetic capacity between actors and the audience. The tendency to disqualify politicians and the disaffection with the "political class" in general is probably one of the causes of the crisis of the model that many observers describe.

On the other hand, in this model, it is the politicians-actors who set the discussion framework: what is talked about and what is out of focus. When this Map is far from the Territory (e.g., when crucial topics for the Territory, such as climate change or the technological revolution and the end of the full employment objective, are avoided), or when obsolete frameworks are in play (e.g., the Nation-State as the playing field of politics in a globalized world), the system wobbles. (For another example of Entropy as the gap between the Map and the Territory, see II.2: "Entropy(ies)".)

This polarity can also be traced in Literature. If conceptualization was the jump from Territory to Map, it could be argued that poetry is the reverse jump: the attempt to recover the fragrances of the Territory (the numinous, emotion, melancholy) from the elements of the text-Map.

This polarity is also observed in narrative (Benjamin, 1936). Chronicle/history is supposedly objective information, generally with explanations that justify the causes of events: the Map of an event. In contrast, the story is a thought or shared experience and involves the narrator. It is more a personal experience (which could be seen and narrated from another point of view), does not claim to be the absolute truth, and does not need to be accompanied by rational explanations: in short, the Territory of an event.

A good representation of the M/T polarity can be seen in the Settembrini/Naphta duo in *The Magic Mountain* by Thomas Mann (Mann, 1924). Settembrini would represent the yin/Territory element: contextualizing, empathetic, diffusely progressive. Naphta is radicality (communist and/or reactionary), incisive argumentation, harsh analysis, and efficiency: the pure Map. Echoes of this polarity are already perceived in the Quixote and Sancho binomial.

In painting, a similar transition between Territory and Map can be observed (Kandel, 2012). The old masters painted "what they knew" (and medieval painting has a strong symbolic charge: white as an image of purity, the lion as a symbol of power, the palm as a symbol of martyrdom, etc.). The discovery of perspective launches a race towards the Territory: the visual arts try to reflect three-dimensional reality as faithfully as perceived. This period lasts about three centuries until photography appears, which occupies that "ecological niche," forcing painting to seek other ways to reflect essential and deep characteristics, not necessarily immediately apparent. The impressionist school tries to capture the fleetingness of the moment, the luminosity of certain outdoor spaces. Cezanne and the post-impressionists reduce volumes to essential bodies and destroy the classic visual perception. Cubism continues this deconstruction work, trying to reflect essences instead of appearances. The expressionist school aims to reflect primary emotions (eroticism, anguish, fear) rather than making realistic photographs. Arguably, these are all attempts to extract a Map from the Territory, thus capturing certain deep characteristics submerged behind the superficial visual aspect.

In music, the Map/Territory polarity corresponds well with Nietzsche's Apollonian/Dionysian dichotomy (Nietzsche, 1872). Thus, J.S. Bach's *Art of Fugue* would be much closer to the Map world, while Carl Orff's *Carmina Burana* would be more representative of the Territory.

Roman (Civil) Law typifies the Map mentality (precise written laws to adhere to, with little room for interpretation). In contrast, Anglo-American Common Law is much more contextual and dependent on the specific interpretation of each case.

In engineering, this problem is central to cyber-physical systems, which integrate the world of physical (analog) machinery and computer systems (digital) (for example, "Internet of Things" (IoT)). An archetype is the problem of time management. The computer world handles a discrete Map-time, with "steps" of the system's internal clock, but no time between one step and the next. In the physical-Territory world, time is analog. When the system becomes sufficiently complex and several processes are managed simultaneously (multitasking), it is easy for these times to come into conflict (Ashford Lee, 2017).

The list of examples could be endless, and it looks very much like the search for the Taoist yin/yang polarity. But a particularly important example due to its social repercussions is the biomedical field. The last section of this text is dedicated to it.

Footnote

1. Naturally, the image of fractality at the border reappears: power lies with the People (Territory), but it has to conform to a Constitution (Map), which emanates from and is ratified by the People (Territory), through a system of political parties (Map)...

III. Map and Territory in Biomedical Sciences

Clinical activity is essentially a round trip from the Territory (medical history, physical examination, additional tests) to the Map (Diagnosis) and back to the Territory (Treatment). Each of these steps brings the advantages and dangers that accompany these leaps.

The process begins with the classic Hippocratic questions (or some version of them): "What is happening to you?", "Since when?", "What do you attribute it to?" After the clarifications requested by the doctor, the physical examination is performed, followed by additional tests.

The next step is a leap to the Map. Based on the collected data, an attempt is made to establish a Diagnosis (category: Map-world). A differential diagnosis is carried out, always in Map-language: discrete, with instruments of the logic of the excluded middle (sensitivity, specificity, etc.: the disease is present or not present, there is no other possibility).

Finally, once the diagnosis is reached, it jumps back to the Territory and Treatment is established. Here reality becomes continuous again, and contextualization is necessary (doses, practicability, adherence...).

Therefore, Medicine is a rich field in which to explore the Map/Territory polarity.

The next chapters will be dedicated to analyzing some of the problems posed by this model. We will see how the choice of conceptual Map decisively conditions clinical practice and allows or prevents seeing certain "ways of becoming ill."

III.1: Medical Nosology

(the branch of medical science dealing with the classification of diseases)

In few areas is the difference between the Map (the Disease) and the Territory (what the patient suffers from) so evident as it is in Medicine. Here, too, the distance between Map and Territory determines the uncertainty – in a way, the entropy – that accompanies medical practice, which is one of its defining characteristics. And although the old adage "there are no diseases, only patients" is almost a cliché, we cannot conceive of clinical practice without the cartography of diseases.

However, an objective analysis of the idea of disease quickly reveals the gap that exists between the disease-Map and the patient-Territory, and shows how our "scientific" idea of disease is severely burdened by *a priori* assumptions inherited from our history.

As cognitive psychology demonstrates, our system of categories (our "Map") decisively conditions our perceptions, and thus we continue to cling to the idea that Disease-entities exist, which we believe to be objective and which guide our clinical practice. Below, some aspects of the "conventional cartography" of Medicine are explored, exposing its dysfunctions and contradictions. Other alternative "cartographic models" will also be explored. Naturally, they will still be Maps, but we will see that to the extent that they approach the Territory differently, they offer new points of view and facilitate the interpretation of certain phenomena.

Conceptual Heritage

Immediately beneath the scientific façade of the classic clinical Disease-entity lies an archaic, quasi-religious idea of disease as possession. A person *is healthy* until, at a certain moment, they *contract (are attacked by)* a disease and become a *patient*. *Nature* may allow them to *overcome* the disease, or it may require the intervention of a *shaman-doctor* to *cure* them. This intervention can be successful (cure), or not (chronic disease or fatal outcome). The term itself (dis-ease) is illustrative, and similar names may be found in many languages.

It is evidently a bellicose account, often simplistic, and we talk about therapeutic *arsenal*, vaccination *campaigns*, reduced *defenses*, etc. (Varela, 1991). All italicized terms fit into that war narrative. As we will see, this approach is not the only possible one (Vargas & Varela, 2013).

Boundary Problems

This conventional narrative is plagued with problems, reflecting the distortion required for the necessary "digitization" to conceptualize (draw the Map). These dysfunctions will be more pronounced the greater the distance between Map and Territory (and therefore the more uncertainty/"entropy" it involves).

Some of the most evident distortions:

- The boundaries between health and disease are often arbitrary and artificial. For example, a patient is diabetic if at any time he/her has a fasting glucose (or glycated hemoglobin) above an arbitrary limit, which varies over the years and according to scientific societies.
- Temporal boundaries are also blurry. At what point does one "acquire" pneumonia? When does one "recover"?
- The differences between syndromes (a set of signs and symptoms) and entities fade. Thus, many entities (e.g., hypertension, diabetes) can be a symptom of another condition (e.g., adrenal tumor or steroid treatment) or a "primary" disease. There is no conceptual boundary

between asthma/disease and bronchospasm/symptom frequently associated with a trivial respiratory infection.

- Senescence becomes an accumulation of "anomalies" (osteoarthritis, cataracts, atheromatosis, etc.) that everyone experiences. Nobody dies of "natural causes" anymore: everyone dies of a disease.
- There are significant discrepancies between the "diseases of doctors," which must be "inculcated" into the culture of patients (obesity, smoking, hypertension...) and the "diseases of patients," often trivialized or ignored by doctors (irritable bowel, fibromyalgia...).
- This situation sometimes reaches delirious extremes. Certain clinical pictures (e.g., vasculitis) have a classification system so removed from the Territory that most patients with these entities are unclassifiable (and therefore not publishable). Only well-labeled (and rare) cases are published, leading to the maintenance of the academic *status quo*, and to an increasingly "gray rheumatology," outside the manuals.

The Concept of Lesion: Anatomoclinical School

A central concept in conventional medical cartography is the idea of lesion. The disease would be the consequence of a lesion on a specific organ or system.

This school of thought originates from the anatomoclinical school of the 18th century, centered on necropsy studies (Lain Entralgo, 1982). The generalization of autopsy propelled unprecedented advances in medical science, and fomented a rational and empirical approach that forever changed the practice of Medicine.

From the beginning, this caused problems in cases where a lesion could not be found in the autopsy, but the idea of a lesional substrate is so powerful that we have no problems accepting "biochemical lesions" (e.g., hyperkalemia) or genetic lesions (e.g., mutations).

The Idea of Cause

Anatomoclinical thinking already had a certain causal character, but this was enormously enhanced with the arrival of Microbiology. Firstly, agents which could be blamed for the disease were found. But above all, with the appearance of antisepsis and antibiotics, the efficacy of medical intervention reached previously unimaginable levels.

This efficacy is what has allowed Medicine to elude (for now?) the crisis of strict causal thinking that most Natural Sciences experienced in the transition from the 19th to the 20th century (quantum mechanics, Heisenberg's uncertainty principle, etc.) (Bronowski, 1978). However, this causal conception is naive and is forced to oversimplify or omit more and more details. For example:

- Less than 10% of patients infected with Mycobacterium tuberculosis develop the disease (Fitzgerald et al., 2020). This forces an explanation of why the vast majority of patients are not affected, and it is necessary to resort to other co-causes (similar to the proegumenal causes of Galenic medicine).
- Many viruses (e.g., herpesvirus) infect the majority of the population at some point. They can insert themselves into the host's genome and remain there indefinitely until some other co-cause (e.g., immunosuppression) facilitates their reactivation. Even this formulation (a "foreign" genome invading "our" genome that reactivates) is probably a projection of the conventional causal prejudice and the conception of self/inside versus non-self/outside.

The Physiopathological School

In the mid-19th century, especially in Germany, a new "medical cartography" emerged, rejecting the disease as a real entity and positing that the lesion is not the cause but the consequence of the pathological process (Faber, 1930). It postulates that clinical pictures should be described and analyzed as quantitative changes of exactly the same processes that operate in the healthy subject. Thus, the disease as ontology or categorical change disappears, and the difference between health and disease is blurred. This school developed a whole series of physiological techniques (thermometry, graphic records, spirometry, etc.), perfectly integrated into conventional clinical practice (although almost always perversely dichotomized: febrile/afebrile, hypertensive/ normotensive, etc.).

However, where the physiopathological cartography triumphed incontestably was in basic sciences and in biomedical research in general. On the other hand, we clinicians have not been able to abandon the classical ontological conception, and we continue to diagnose and cure diseases. This has been a source of increasing discord. Almost all relevant biomedical advances occur in "physiopathological language" (continuous, anti-ontological), which clinicians have to read and translate into our "language of discrete clinical entities." This task of reading and translation is increasingly painful (Lenfant, 2003), and it is undeniable that patient management will increasingly depend on a map not based on "clinical entities" (Joe, 2022; Swerdlow et al., 2016).

The Deserters of the Classical Conception

It is striking how some of the subspecialties emerging from the classical causal idea have become its worst enemies.

Immunology

An example is Immunology, which was born under the auspices of Microbiology and aimed to explain the specificity of an individual's response to the "attack" of a germ (Silverstein, 1989). This fits perfectly with the classical concept of disease as a well-defined entity with a clearly identifiable cause, giving rise to a successful police metaphor of "chasing invaders." Additionally, this scheme resonated with the intuitively very attractive idea of "self versus non-self."

However, a variant of "inadequate" immune response soon emerged, which damaged the patient instead of protecting them. These types of reactions were called allergic and gave rise to an entire specialty (Allergology), which, although uncomfortable for the conventional paradigm, did not change it. Even more uncomfortable were autoimmune diseases, where certain clones directly attack the body. Still, the general scheme was maintained, explaining the phenomenon as "evil or mistaken policemen," without questioning the model.

However, this scheme is now difficult to reconcile with the current view of the Immune System as an idiotype/anti-idiotype network offered by basic research (Burdette & Schwartz, 1987). In this model, the immune system is conceived as a large, interconnected network, where each immune response (idiotype) triggers an anti-idiotype response (and this perhaps another, etc.), forming a mega-structure in continuous fluid equilibrium, like the surface of a perpetually moving lake, where the conception of "police versus invaders" is at best naive.

These are not theoretical fantasies and have immediate clinical consequences. For example, it is known that in bone marrow transplants for certain hematological tumors, there is no "replacement" of one immune system by another, and there is always a balance/struggle between the original immune system (host) and the grafted one. Excessive predominance of the original facilitates tumor recurrence; excessive predominance of the grafted one aggravates the symptoms of graft-versus-host disease (Weiden et al., 1981). Similarly, a treatment increasingly used in certain immune pathologies is the administration of massive doses of nonspecific immunoglobulins, as a "reset" of the immune system (Ballow & Shehata, 2022).

These findings are perfectly explainable in the idiotype/anti-idiotype network model but are difficult to justify in the conventional "police" model.

In summary, the original Map of quantized, specific immune response to a particular "foreign agent" is replaced by a fluid, continuous model, where "normal" and "pathological" cannot be cleanly separated, much more in line with physiopathological thinking.

Genetics

A similar drift can be observed with Genetics. Initially, Genetics seemed to offer an ideal fit to the conventional pathological model. The genetic disease was due to a mutation, which was either present or not (there were no "gray areas") and which caused the disease. In short, it was the paradigm of conventional medical cartography: a qualitative change with a clear causal substrate.

It soon became apparent that the causal chain was complicated: how, where, when, and why a mutation occurs is only addressable from a statistical point of view, and the causal explanation of a mutation is usually impossible. This issue was still manageable for classical genetic diseases, which comprised a large number of clinical entities but a very small number of patients.

However, it soon became obvious that many extremely common pathologies (hypertension, diabetes, obesity, etc.) had a strong genetic load, but did not follow the conventional Mendelian inheritance pattern. The issue became even more intractable when faced with somatic genetic diseases (mainly tumor diseases).

Moreover, the initially clean boundary of "mutation -> disease" quickly became blurred:

- The concept of penetrance had to be introduced, accounting for the proportion of cases with the mutation that did not develop the disease.
- The mutation was not a dichotomous datum. For example, in certain neurodegenerative diseases (e.g., Huntington's chorea), the appearance and severity of the clinical presentation depended on the number of repeated triplets (a quantitative, rather than qualitative, variable).
- The same disease could be caused by completely different mutations (e.g., elliptocytosis or Charcot-Marie-Tooth disease).
- The appearance of the clinical presentation could be decisively conditioned by certain environmental factors (e.g., in porphyria or hemochromatosis) or by other co-modifying genes (e.g., congenital deafness).
- There is a mitochondrial inheritance, derived only from the mother, which does not fit the classic Mendelian pattern.
- The conventional concept of Genetics postulated that information flows unidirectionally, from DNA (genotype) to proteins (responsible for the phenotype). It was assumed that any information in the reverse direction (phenotype -> genome) passed through a mutation and a process of Darwinian natural selection. However, this idea is also entering into crisis. There is data showing a certain phenotypic plasticity and suggesting that the environment can modify the transcription process of the genome without involving mutations or going through the evolutionary bottleneck.

These and other causes made Genetics, born under the umbrella of classical nosology (i.e. the classification of diseases, with a clear separation between health and disease and a recognizable cause), abandon the classical Map and become a perfect field for the physiopathological model: continuous, without lines separating health/disease, and a perfect model for systemic thinking.

III.2: Medical Narratives

A Map is a selective representation in which the cartographer has decided which aspects to expose and which to omit. Therefore, there are physical, political, climate maps... Each Map entails a narrative. It is obvious that the Map one has decisively conditions what one finds: you only see what you know. Therefore, the "narrative" of diseases determines not only how they are classified and ordered, but also what actions are taken or omitted.

War Narrative

In Medicine, the classic narrative is the war discourse, the fight against a disease/morbid entity, often with a well-identifiable causal agent. This perfectly aligns with the Judeo-Christian mentality of guilt and the primordial metaphor of the inside/outside or me/not-me (Lakoff & Johnson, 1999). This narrative permeates the entire daily practice of medicine, and especially the general perception of the patients themselves. It is telling that during the periodic briefings given by the Spanish Government during the Covid-19 epidemic, the technical manager was often surrounded by the Minister of the Interior and two generals...

However, apart from a certain simplism, this narrative gravely distorts our understanding of the pathological process. For example, this preconceived notion of "war against the virus" has delayed the realization that Covid-19 disease is not simply the "attack of a virus." This may summarize the first part of the disease. But there is a second part, of hyperimmune response and macrophage activation syndrome, which is responsible for a significant part of the morbidity and mortality.

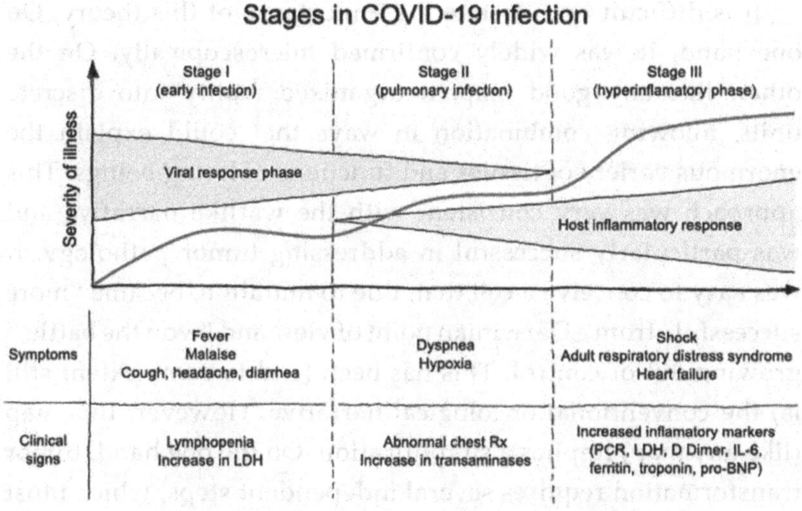

Fig. III.2.1: Phases in Covid-19 infection

This "selective blindness" is not new. In most cases of viral myocarditis, hepatitis, or pancreatic insulitis, the damage is not caused by the virus but by the immune response (Lok, 2022; Cooper Jr., 2009; Lai & Chopra, 2020; Hirsch, 2022). This same selective blindness supports the search for the "broad-spectrum antibiotic" (ideally, of total spectrum) and makes it difficult for us to accept the idea of the microbiome as part of our "self" (Rees et al., 2018).

Cellular Narrative

Another cornerstone of the prevailing medical paradigm is the Cellular Theory, crystallized by Rudolf Virchow in the mid-19th century. Essentially, it establishes that the cell is the "unit of life." There is no life without cells, and the cell constitutes the central axis — structural and metabolic — of all living beings. Pathological processes result from a dysfunction in cellular physiology.

It is difficult to overstate the importance of this theory. On one hand, it was widely confirmed microscopically. On the other, like any good map, it organized reality into discrete units, allowing combination in ways that could explain the enormous variety of tissues and functions of living beings. This approach was very consistent with the warlike narrative and was particularly successful in addressing tumor pathology. It was easy to conceive a cell that, due to mutation, became "more successful" from a Darwinian point of view and "won the battle," growing out of control. This has been (and to some extent still is) the conventional oncological narrative. However, this map (like any map) implies a simplification. On the one hand, tumor transformation requires several independent steps, which must be completed one by one (Hanahan & Weinberg, 2000):

1. Becoming independent of growth factors, usually necessary for cell reproduction. This requires some mutation that permanently activates the receptor or the signaling pathway (e.g., mutations in HER2 or EGF (epidermal growth factor)).
2. Being able to bypass inhibitory signals (e.g., mutations in the retinoblastoma protein gene pRb or c-myc).
3. Evading apoptosis (programmed cell death) by overexpressing anti-apoptotic genes (e.g., Bcl-2) or inactivating central apoptosis genes (e.g., p53).
4. Acquiring the capacity for unlimited replication. One of the central mechanisms in the control of cell replication is the shortening of telomeres (chromosome ends). Each mitosis loses 50–100 base pairs from the chromosome's end. This allows only a limited number of mitoses (generally, between 60–70 replications). If this limit is exceeded, the chromosomes become unstable, fuse with each other, and the cell dies. Therefore, one of the central mechanisms in tumor physiology is escaping

this mechanism (e.g., mutation of the telomerase gene, responsible for the cuts).

5. Maintaining angiogenesis. When the tumor grows, it needs to be accompanied by vessels that nourish it (which are of another cell lineage). This can be achieved by overexpressing endothelial growth factors (VEGF) or fibroblastic growth factors (FGF).
6. Acquiring the ability to infiltrate and/or metastasize to other tissues. This requires changes in adhesion proteins (cadherins, integrins) and the overexpression of proteolytic genes.
7. A nonessential but very facilitating characteristic is the development of genetic instability, usually due to mutation of DNA editing and correction mechanisms. This obviously decreases growth efficiency (frequent nonviable genetic errors), but greatly facilitates acquiring the various abilities that enable tumor development (and the emergence of chemotherapy resistance).

In summary, the narrative of the cell that mutates and escapes control is somewhat oversimplified. The process is a long path that requires overcoming several successive and independent barriers.

Virchow considered the cellular structure a community, subject to control mechanisms. Tumor transformation could be considered a kind of "plane transition." The tumor would stop belonging to the community (and being subject to its control mechanisms) and become an autonomous individual, operating as an "independent Darwinian being" (Levin, 2019). It no longer ties its fate to the rest of the community, escapes its regulatory systems, and optimizes its evolutionary fate autonomously.

There are parallels between cellular theory and Newtonian physics. Both were revolutions that radically transformed their disciplines; both had such radical success that they became "obvious" and indisputable. And both are reductionist theories, more focused on objects than on relationships. This is one of the main problems faced by the cellular narrative. Inevitably, the cell is immersed in a context that modifies it, sometimes decisively (for example, the cellular context decisively conditions breast cancer behavior (Arendt et al., 2010)). Gradually, oncology is becoming less mechanistic and acquiring a more systemic profile.

Economic Narrative

The war narrative has been typical during major epidemics (one example being the recent Covid-19 pandemic). But above all, it has been the hegemonic narrative in the doctor-patient relationship (understood as doctor-individual with patient-individual). The cellular narrative (which can be harmonized well with the war narrative) remains predominant in Oncology.

However, when it comes to managing a healthcare system composed of many individuals, the war or cellular narratives are insufficient. When resources are limited, and investments in certain fields or population groups come at the expense of others, issues of equity and prioritization arise. There is a need to quantify benefits to compare and choose, and a whole new management language, increasingly sophisticated, and generally derived from the world of economics, is developed. It soon became clear that measuring gross survival was not enough and that it depended on which population and, most importantly, the quality of life. Instruments like QALYs (quality-adjusted life years) are developed, the "output" of hospitals is quantified, their "profitability" is analyzed, the cost-effectiveness of an intervention is assessed, and attempts are made to adjust salaries according to these criteria.

I believe that, at least in my country (Spain), it can be said that physicians, in general, have not embraced this narrative (at least in the Spanish National Health Service). The physicians involved in healthcare management had no choice but to engage, but the bulk of practitioners maintained an attitude of rejection and often distrust towards colleagues who entered that game. Many doctors who joined this narrative soon had to choose between returning to the "arena of their peers" or staying definitively in the world of management (and often being seen as traitors). In any case, the interface between managers and practitioners has never been fluid and has always had the problem of narrative differences.

Naturally, dysfunctions arose. For example, the idea of the market as a way to optimize resources has been stopped in many medical areas (e.g., transplants), but it continues to push softly and constantly.

It is important to note that the economic narrative still relies on the classic dichotomous thinking of healthy/sick. This dichotomy made it easier to identify which population the healthcare system should be responsible for.

A probable scenario is that this "red line" separating health and disease will dissolve, and society will have increasing access to "health improvement" technologies and anti-aging treatments. For example, cochlear implants were typically reserved for childhood deafness. However, recently, their usefulness in age-related deafness has been demonstrated (Carlson, 2020). The cost of this technology, or other "anti-aging" treatments applied to a massive candidate population, may be unsustainable for the healthcare system. It will be very difficult for the conventional war narrative to manage this scenario, and instead, it will be fertile ground for the economic narrative. But this could break the conceptual seams of the healthcare systems as we know them.

Ecological Narrative

Another possible approach to many highly prevalent diseases is to explain them as a mismatch between the pace of genetic

(Darwinian) evolution and environmental conditions. The disease would be a (mal)adaptation of genes to an environment that changes too quickly for genetic evolution to keep up. The examples are countless.

- **Hemochromatosis**
  Hemochromatosis (iron overload disease) is the most prevalent genetic disease of Mendelian inheritance, and is a significant cause of morbidity and mortality, mainly due to the risk of cirrhosis and hepatocellular carcinoma. The mutation appeared in Europe approximately 4000 years ago (Distante et al., 2004) and spread rapidly, now being present (in heterozygosity) in a high percentage of the population (between 2% and 13%, depending on the population and the mutation considered) (Steinberg et al., 2001).

  Why does a "pathological" mutation spread so successfully and quickly around the world? The most plausible explanation is that this mutation, which involves a greater capacity to accumulate iron, is a blessing in a situation of difficult access to iron (especially in women of childbearing age). However, in a society like ours, with easy access to this trace element, the mutation becomes a threat. But even today, in many populations, it is observed that the probability of reaching centenarian ages is significantly higher in women carrying this gene (Carru et al., 2003; Lio et al., 2002).

- **Hypertension in the African American Population** (Roberts, 2001)
  It is well established that the African American population has a much higher incidence of hypertension than the general US population (and Africans). One

possible explanation for this phenomenon is that this population mostly descends from slaves brought from Africa, originating from areas with a low-salt diet. This population had to face a brutal journey on slave trader ships, subjected to severe dehydration and frequent diarrhea. The hypothesis is that this created a strong selection gradient favoring "salt-saving" genetic constitutions. Over time, when this genome encounters a salt-rich environment, the incidence of hypertension skyrockets.

- **Obesity/Diabetes and "Thrifty Genotype"** (Neel, 1999) The high prevalence of obesity and diabetes ("diabesity") is a condition with a high genetic component and significant morbidity and mortality. Its increasing prevalence has been explained by the selection of certain "thrifty" genes in populations subjected to periodic famines or long journeys between islands. These genes would prioritize fat accumulation as a possible energy substrate for future times of scarcity. When these populations move to an environment with easy access to high-calorie diets, this initial evolutionary advantage becomes a burden, with a disproportionate increase in the incidence of obesity and diabetes.

Similar "ecological" explanations can be found for many prevalent diseases. The examples are countless: the incidence of sickle cell disease would be increased in the African American population because the responsible gene offers some protection against malaria; the incidence of tumors is much higher in the elderly because there are genes that delay tumor appearance until after reproductive age (when the individual is "out of the Darwinian game"); the diseases that decimated indigenous populations during the American colonization are explained by contact with pathogens to which the colonizers had been exposed for centuries but were new to the natives, etc.

In each case, it is a dissociation between the rate of genetic evolution and the rate of environmental change. In short, a distancing between the Map (genetic) and the Territory (environment in which it moves).

This change of perspective allows for new alternatives, where the concept of Information completely changes the rules of the game. Thus, conceiving the Map (genome) strictly as Information allows for its modification in real-time (e.g., by gene editing mechanisms like CRISPR editing (https://en.wikipedia.org/wiki/CRISPR_gene_editing), making genetic evolution go from Darwinian to Lamarckian).

Obviously, this narrative does not harmonize well with the classic "war" thinking, nor with conventional lesion language. However, it connects perfectly with the language of basic sciences and physiopathological approach (see III.1: "Medical Nosology").

III.3: Statistics

Another clear border post between the Map and the Territory is Statistics. Essentially, it is a tool that aims to detect deep patterns and rules, eliminating the potential circumstantial component. In summary, it aims to identify the Map underlying the Territory by removing the "noise" of contingencies. Its usefulness is evident and underlies almost all significant medical advances. It is the conceptual basis on which Evidence-Based Medicine rests (see III.4: "Evidence-Based Medicine").

Like many other transitions from Territory to Map, it involves converting an analog reality into a discrete, generally dichotomous image (e.g., "males have higher average creatinine levels than females," or "there is a direct correlation between weight and creatinine levels," or "oxygen therapy is an effective treatment for pneumonia"). As we will see, this transformation from quantitative to qualitative lies behind some of the critical problems of statistics. Moreover, this seemingly cold and objective leap from Territory to Map is laden with assumptions, of which users are not always aware, and which explain some of its famous paradoxes and errors:

1. Statistics does not "prove" anything: it only serves to reject the null hypothesis; that is, it estimates the probability that the data obtained (e.g., differences between populations or correlation between variables) are due to chance (and therefore do not represent any real difference or correlation). This is the meaning of the p-value that accompanies every statistical test ($p<0.05$ means that the probability that the results obtained are due to chance is less than 5%). For example, we can ask whether the average height or cholesterol levels differ

between women and men. In the attached table, the average height and cholesterol levels of a population of women and men are shown. In a sample of 262 subjects (Varela et al., 2016), the following results are obtained:

Table III.3.1

	Men (mean, stand. dev.)	Women (mean, stand. dev.)	p
Height (cm)	170.2 (6.4)	156.7 (6.9)	<0.001
Cholesterol (mg/dl)	187.8 (28.4)	180.5 (27.8)	0.06

The p-value <0.001 in the height comparison indicates that in a new sampling of that population, the probability that men will not have a higher average height than women is less than 1 in 1000. On the other hand, we can say that the probabilities that the average cholesterol is higher in men than in women are greater than 90%, but not 95%. It could be stated that there are significant differences in cholesterol between sexes if we accept a confidence level of 90%, but we cannot assert this if our required confidence level is 95%.

Therefore, the results of the "Map" (whether or not there are significant differences between genders) will depend on the confidence level with which we want to make the assertion (p-value), and this is completely arbitrary. Additionally, the argument only works in one direction (demonstrating differences). On the contrary, we could not say that "treatment X is not effective for disease Z." We can only say that "we cannot affirm that X is effective because we have not detected significant differences (with the established threshold) between treated and untreated patients."

2. It is based on an ontological conception of disease. As with any transition from Territory to Map, the first step is to "discretize" the object of study. The essential concepts of medical statistics (Sensitivity, Specificity, Predictive Power, etc.) (Carrasco, 1989; Sackett et al., 1996) are based on an excluded middle logic: the disease is either present or not present. This is a pixelation that we tend to accept without problems, but besides the boundary issues (when does it start? when does it end?), it implies uncritical acceptance of a conception of disease as a qualitative change, which is fraught with problems (see section III.1: "Medical Nosology").

3. Simpson's Paradox. An example of the risks involved in the transition from quantitative to qualitative that accompanies the shift from Territory to Map is Simpson's paradox. Let's imagine a sample of 390 subjects (fictional data). The aim is to analyze the influence of alcohol and tobacco on the risk of ischemic heart disease.

Table III.3.2

	Smoker	Non-smoker	Total
Ischemic heart disease (IHD)	55	70	125
No Ischemic heart disease (No-IHD)	25	240	265
Total	80	310	390

The incidence of ischemic heart disease is 68.7% in smokers (55/80), compared to 22.6% in non-smokers (70/310): there is an association between tobacco and ischemic heart disease.

Let's now look at the results for alcohol use:

Table III.3.3

	Drinker	Non-drinker	Total
Ischemic heart disease (IHD)	60	65	125
No Ischemic heart disease (No-IHD)	100	165	265
Total	160	230	390

The incidence of ischemic heart disease is 37.5% among drinkers (60/160), compared to 28.2% among non-drinkers (65/230): there is also an association between alcohol and ischemic heart disease.

The obvious conclusion would be that both alcohol and tobacco are harmful to the heart.

The problem arises when analyzing separately the populations of smokers and non-smokers:

Table III.3.4

	Smokers		Total	Non-smokers		Total
	Drinkers	Non-drinkers		Drinkers	Non-drinkers	
IHD	40	15	55	20	50	70
Non-IHD	20	5	25	80	160	265
Total	60	20	80	100	210	310

In smokers, alcohol decreases the risk of ischemic heart disease (66% (40/60) in drinkers, versus 75% (15/20) in non-drinkers). But even in non-smokers, alcohol also decreases the risk of ischemic heart disease! (20% in drinkers (20/100), compared to 24% in non-drinkers (50/210).) This seems quite counterintuitive, but it becomes clear when analyzing the distribution of the sample. The incidence of smoking among drinkers is much higher than among non-drinkers (37% versus 9%). In summary, alcohol has a certain protective effect (less potent than the harmful effect of tobacco), but there is a clear association between alcohol and tobacco use (drinkers are more likely to be smokers), which masks the protective effect of alcohol. The inherent "pixelation" of making a Map hides the detail of the irregular distribution of subpopulations and reverses the real effect of alcohol on ischemic heart disease.

4. Statistics implies a conceptual leap from sample to population. Another not always conscious (and often difficult to verify) assumption is that the sample from which conclusions are drawn (the Map we want to obtain) sufficiently represents the population being discussed. An example of this type of error was the initial data on hormone replacement therapy (HRT) in postmenopausal women (van den Beld & Lamberts, 2021). The first observational studies, comparing women who had or had not received HRT, showed clear advantages for the women treated (versus the untreated). However, a randomized study (Rossouw et al., 2002) (where inclusion in the treatment or placebo arm was random) showed that women receiving HRT had a higher risk of cardiovascular disease and gynecological tumors. The explanation for this paradox was that, in the first (observational) studies, women starting HRT already had a better survival profile at entry (younger, higher social level, better healthcare coverage). The problem with Maps obtained from observational studies is that one has to decide *a priori* which variables to adjust for (and this is a subjective and debatable decision). Controlled studies attempt to overcome this problem by relying on the distribution of unconsidered variables to be random and representative of the real distribution in the population.

III.4: Evidence-Based Medicine

Another clear boundary post, fault line, or bridge between Map and Territory is the relationship between the school of Evidence-Based Medicine and daily clinical practice.

The school of Evidence-Based Medicine (EBM) was born as a response to the great variability in medical practice (both in techniques and results) in different settings, and as a reaction to teaching based on opinions issued from a position of "authority," often without experimental support. Its assumptions were quite indisputable: it was demanded that any clinical activity (diagnostic or therapeutic) be based on established facts, and it was proposed that for each reasonable doubt, answers should be sought by analyzing published papers. To do this, it offered relatively objective criteria regarding the quality of the evidence found.

These basic principles (Sackett et al., 1996) were quite unquestionable. Many basic scientists were even surprised by the name of the movement ("if not on evidence, on what else could medicine be based?"). No reasonable person could oppose this plan.

However, soon problems arose that, without questioning its core, compromised its practice.

One of the first problems came from economic managers, who argued (especially for very expensive treatments) that the public health system should not cover the costs of treatments for which there was no "evidence." This was very serious: there are rare pathologies for which it is impossible to gather a sufficient sample to conduct a formal trial. In these cases, actions are based on physiological plausibility criteria, reasoning by analogy, or isolated, unsystematizable experiences. But not offering a potentially curative treatment because sufficient evidence has not been gathered was very contentious.

The second practical problem was efficiency. Seeking evidence to answer each of the clinical doubts that arise in daily practice required time and dedication that was hardly compatible with the usual clinical pace. And the idea of promptly answering each clinical question by seeking evidence became an ideal to strive for rather than a real practice. Finally, with the arrival of Clinical Guidelines (see below), this term became a crutch to avoid thinking.

Supposedly, EBM is a fight against all apriorism, asking for an objective evidence endorsement for any statement and seeming immune to any bias. However, it drags along with it some inherent concepts of the "Map" that supports it, which decisively conditions it:

1. It inherits an ontological conception of disease. The patient can either have or not have the disease (regardless of whether we know it). This is a presupposition: Aristotelian logic of the excluded middle. This is the basis of all the instruments EBM handles (sensitivity, specificity, predictive power, test power, etc.). This discipline can handle continuums in terms of probability, degree of uncertainty, etc. But disease is a categorical, not continuous, entity: an individual is either afflicted or not afflicted by it.
2. While seemingly unbiassed, the "search for evidence" is entirely loaded with apriorisms: most notably, which factors are considered relevant and which may be omitted. For example: should a 72-year-old patient with chronic atrial fibrillation be anticoagulated? The immediate answer is: yes. But he has severe arterial hypertension,

which increases his risk of cerebral hemorrhage (we can include this in the evidence search criteria). He had a digestive hemorrhage 5 years ago (Do we know the risk of recurrence? Do we include it as a variable in the evidence search? Is the degree of hemodynamic repercussion or the etiology of the hemorrhage important?). He has a certain visual deficit (How much? Do we include it?) and lives alone in an apartment with stairs, living with several cats (Should we omit this?: it may entrain considerable risk of falls…). Additionally, he is diabetic (we could include it in the search) and has been diagnosed with COPD (Do we include it?). When seeking evidence for this case, it is crucial which criteria are included. Naturally, the more criteria included, the more adjusted the search will be, but the less likely it is to find robust studies to answer the question.

In short, under the guise of objectivity, Evidence-Based Medicine carries an entire conception of disease and leaves to the user's discretion which factors are considered relevant when seeking specific evidence.

Clinical Guidelines

One of the corollaries of EBM has been a number of Clinical Guidelines, which theoretically collect the best evidence to optimize the management of typical cases. As an attempt to spread best practices and reduce variability, it seemed an impeccable initiative. However, apart from the problems previously mentioned, at least two more soon emerged:

1. They became "shortcuts to avoid thinking," authentic catechisms. Once the patient was framed into a "typical case," it was enough to apply the corresponding Guide. This became even clearer when Clinical Guidelines began

to be used as judicial criteria in medical claim cases: the way to avoid problems was to follow the Clinical Guide and not give it more thought. In summary: better to be guided by a (bad) Map than risk getting lost in the Territory.

2. Obviously, Clinical Guidelines became the star target of the pharmaceutical industry. It is impossible to prove they have been infiltrated (and I suspect that, in most cases, they have not (yet)), but in a context where the State has relinquished the initiative of research and postgraduate medical training to the pharmaceutical industry, I think the future is gloomy. Certainly, the initiative of clinical trials lies entirely with the industry, and the bulk of the "evidence production" is done along the paths it marks. In summary, it is the industry that is building the Map. For those of us who are skeptical about the absolute goodness of the Market in any situation, this scares us.

III.5: Other Possible Maps. Fractal Pathology

We tend to conceive of maps as spatial objects, representing relations of physical proximity. However, some of our most elaborate cartographic constructions do not operate with space. One of the most obvious examples is the genome. It is the fundamental map/guide responsible for the synthesis of all the proteins that will eventually constitute the Territory of each individual. But it is not the only example.

Maps in Time

An important family of maps is not established in space but in time. Or more precisely, in frequency.

Vision

An example is chromatic perception. Our "color map" represents differences in the frequency of the light (electromagnetic wave) that hits the eye (from 400 THz for red to 790 for purple) (Fig. III.5.1). Each color tone corresponds to a vibration frequency. Finally, the categorization into colors is another example of the leap from Territory (continuous) to Map (discrete), which greatly enhances the capacity for visual recognition.

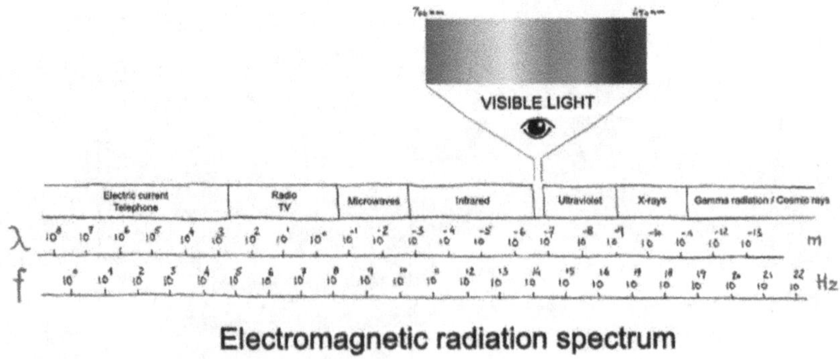

Fig. III.5.1: Frequency spectrum of electromagnetic radiation

Hearing

Sound is another example of a frequency map, in this case of pressure waves, with a frequency ranging from 20 to 20,000 Hz. The wave trains are transmitted through the eardrum and the chain of bones in the middle ear to the cochlea.

In this case, there is also a certain spatial map (Fig. III.5.2): the cochlea has an ordered distribution of neurons more sensitive to certain frequencies, thereby achieving the equivalent of a Fourier transform (see II.7: "Fourier Transform"), creating a frequency map and thus "digitizing" the wave trains corresponding to auditory stimuli.

This allows them to be used as Map-objects, with all the consequent advantages: we can conceptualize phonemes (and construct languages), tones (and make music), sounds (and recognize the barking of a dog), etc.

A difference between the frequency maps of vision (electromagnetic radiation) and hearing (pressure waves) is that the Fourier transform of hearing allows us to discriminate frequencies (and we can recognize a piano and a violin playing the same note in unison, or the three notes that form a chord). On the other hand, the eye only recognizes the resulting frequency, and we are not able to distinguish a pure green color from a combination of blue and yellow. This lower discrimination is, however, compensated for by much greater spatial precision.

Nerve Impulse

The nerve stimulus itself is encoded as frequency. The intensity of response of a nerve ending is not encoded as intensity in the discharges of action potentials but as an increase in the frequency of these potentials.

Bioelectric Networks

A paradigmatic example of a Map in biology is the genome. It encodes all the structural and regulatory proteins of the

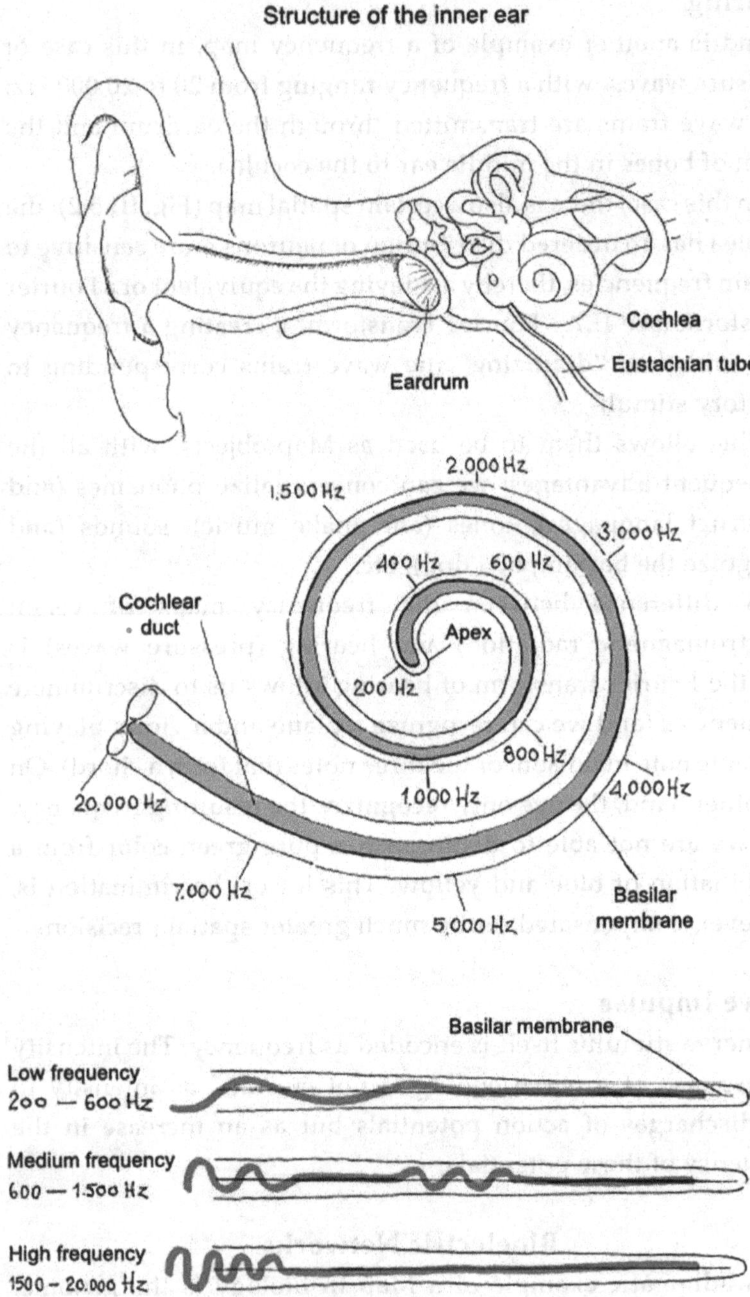

Fig. III.5.2: Structure of the inner ear

organism and is an archetype of encapsulated information. The conventional Darwinian model posits that the genetic Map manifests as phenotypic Territory, and natural selection acts on this.

It is important to consider that natural selection operates on the Territory, not the Map. This leap from genotype to phenotype (in essence, a paradigmatic example of the leap from Map to Territory) is the domain where Epigenetics operates. In addition to conventional mechanisms (DNA methylation, histone modification, non-coding RNAs, etc.), there is another Map that could be particularly relevant in this leap, especially in mechanisms of morphogenesis (the growth and organization of anatomical structures) and differentiation/dedifferentiation (including oncogenesis): bioelectric networks.

One of the most typical characteristics of any cell is the presence of a membrane potential (Hall & Hall, 2021), that is, a voltage difference between the inside and outside of the cell. This is maintained by a sophisticated mechanism of ion channels that traverse the membrane and have a selective capacity to allow or prevent the passage of certain ions. This leads to different concentrations of ions and electric charges between the intra- and extracellular media. These gradients are used differently and for various purposes in different organs and tissues.

Recently, it has been observed that these membrane potentials are not homogeneous, and there is a gradient of membrane potentials within different tissues. This Map of membrane potentials forms a bioelectric network and constitutes a layer of information that crucially conditions morphogenesis. Thus, strictly physicochemical manipulations that modify the bioelectric network (without affecting the genetic map) can generate an eye on a tadpole's abdomen or a head on a planarian's tail (Levin, 2021).

This bioelectric network constitutes a new Map of information that, while not completely independent of the genome (as it requires the existence of ion channels based on proteins encoded by the genome), is capable of acting "on top" of the genetic information, conditioning the "leap from Map to Territory" characteristic of the phenotypic expression of a given genome. This role as an intermediate layer between genotype and phenotype has important advantages: it is a structure that allows a certain modularity and facilitates a low-cost exploration of possible "evolutionary spaces."

Bioelectric networks have some special characteristics:

- Unlike the genetic map, they are easy to reprogram over and over again.
- Their stability over time gives them the capacity for memory and computation.
- They do not have a direct correlation with the genome. On the one hand, similar bioelectric configurations can be achieved with different arrangements of ion channels. On the other hand, the same genetic coding can result in different bioelectric configurations, conditioned by the environment.

In essence, this is not a newcomer in biology. This capacity to integrate information in space and time has been used for millions of years (e.g., in bacterial biofilms (Prindle et al., 2015) or in morphogenetic mechanisms (Levin, 2023)). Arguably, a mechanism that is both metabolically "expensive" and ubiquitous should have evolutionary relevance. In fact, the neuron can be interpreted as an optimization of the bioelectric network system both in space (axons) and in time (conduction speed, myelination). Thus, a function that initially operates in 3D physical space (e.g., folding of embryonic layers or focal growths in embryogenesis or morphogenesis) evolves to operate in elementary motor functions (chemotaxis, phototaxis), more complex motor functions (muscular system), until reaching cognitive functions that define consciousness. Essentially, behavior differs from morphogenesis mainly in the range of times over which it operates.

Structural Layers

One of the most striking characteristics of Life, which tends to go unnoticed because it is so evident, is its tendency to organize into superimposed organizational layers: metabolic networks are organized within the cell, cells form tissues, tissues form organs, organs are arranged in systems, systems are part of individuals… The most efficient way to describe each of these layers depends on the level considered: each level has its own "Map." The tools we use to analyze the cellular level do not contradict but are not useful for describing the anatomical or behavioral level. We accept the existence of the B lymphocyte and the tiger, but they correspond to different maps.

What does it mean to exist? The idea of how existence could be defined as maintaining a differentiation between what "is" and the exterior, and being able to maintain this differentiation over time, was discussed in I.3: "Mind/Map versus Brain/Territory." It was also shown how the Free Energy Principle (Friston, 2009, 2010; Friston et al., 2010) allows this relationship to be formalized, pushing towards the development of predictive processing, where survival is directly related to the match between the system's predictions and the afferences it receives. The classic model is the T.O.T.E. cycle ("test – operate – test – exit") by which elementary homeostatic mechanisms function. This approach allows the behavior of individuals to be described as "pursuing a goal" (minimizing their free energy = optimizing their survival) and permits behavior to be analyzed as "goal-seeking" without falling into teleonomy.

One way to optimize survival is to improve predictions, and a clear path is improving mechanisms for sensing/responding to the environment and refining conjectures in time, space, and causal chains. This would push towards differentiation and specialization of some parts of the "being" in question. It is not difficult to see how this staircase would lead to the appearance of membranes (and the cell), the differentiation of nucleus and organelles (eukaryotic cell), the differentiation of functions (tissues, organs), systems (central nervous system, locomotor system, etc.).

This pressure to optimize predictions underlies "infotaxis," the drive towards information, which increases as the "being" becomes more complex. This is probably the mechanism underlying the irresistible tendency of the Geist formulated by Negarestani (Negarestani, 2018) to "rationalize" its entire environment.

In this context of sensing/acting on the environment, Michael Levin proposes an evolutionary model of cognition (Scale-Free Cognition) (Levin, 2019) that postulates a progressive expansion of the scope that can be sensed and acted upon by the "being." Graphically, if time is represented on the vertical axis and space on the horizontal axis, the cognitive capacity of an individual would be represented by a double cone, showing its ability to foresee/act on its environment (horizontal axis) and be influenced by its past and act on its future (vertical axis) (Fig. III.5.3A).

This allows for a gradualist view of cognition and avoids the anthropocentric bias of intelligence as something that appears abruptly and is exclusive to humans. In Levin's model, perhaps the only thing that characterizes the human being (for now) (Fig. III.5.3.B) is that their ability to foresee time is several orders of magnitude greater than their life expectancy (essentially, a reformulation of the classic thesis that points to the consciousness of death as the fundamental difference

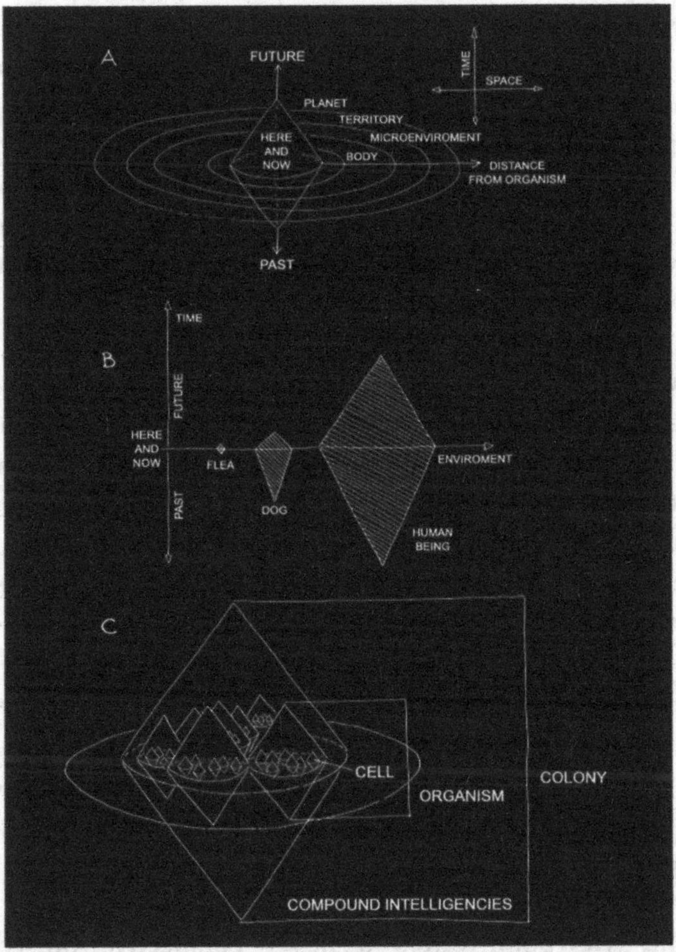

Fig. III.5.3: A. Cognitive cone of each being. On the horizontal axis, the space/distance over which it can act. On the vertical axis, the time it can remember or foresee/act upon. B. The "cognitive cones" are different for different species, both in time and space. C. "Cones within cones" can exist. Drawing modified from Levin, M. (2019). The Computational Boundary of a "Self": Developmental Bioelectricity Drives Multicellularity and Scale-Free Cognition. *Frontiers in Psychology*, 10. https://doi.org/10.3389/fpsyg.2019.02688, under Creative Commons license.

between humans and the rest of living beings). It is also clear that there can be "cones within cones" (e.g., mitochondria within neurons within a central nervous system within a living being) (Fig. III.5.3.C). Each step increases the "cognitive cone" and improves the subsystems' survival. The closer we are to the "here and now," the more we interpret a response as "automatic." The further we move away from this point, the actions will seem more "voluntary" or "intentional."

Spaces of Action

If we define a "space of action" as the set of possible situations over which the individual can know/influence (what we might call the counterfactuals), it is clear that the more complex the individual, the larger this space of action will be.

It is also clear that the action space of a subsystem is deformed by the higher system that encompasses it. It is not that the subsystem no longer "pursues its selfish ends"; it is rather that, being included in the higher level, its "selfish ends" are modified (= its action space is deformed). For example, the sexual instinct or maternal instinct alter the action space of an individual (they are metabolically very expensive and can compromise its existence) and only make sense to the extent that it is part of a species.

This viewpoint offers a novel perspective on certain pathologies. For example, tumor development could be interpreted as a loss of control at the macro level, with the subsystem navigating its own "action space," not "deformed" by the higher plane. This independence results in an initial liberation but reduces its survival (obvious in cancer). In this sense, it is noteworthy how tumors may develop by isolating a cell (Yamasaki, 1995), how the environment can restore a cell to normality (Kasemeier-Kulesa et al., 2008), or the finding of a characteristic loss of connections between cells in oncogenesis (Knights et al., 2012; Zhou et al., 2023).

Fractal Pathology

Fractal structures are ubiquitous in the organism. In general, they are used as "dimensional exchangers," when it is necessary to connect a transport structure ("pipetube": 3D) with a diffusion or absorption structure ("surface": 2D). Obvious examples are the capillary system, the bronchial tree, the intestinal mucosa, the portal systems, the cardiac conduction tissue, and many others.

However, our classical paradigm has few tools to analyze these structures. Classical geometry operates in two perfectly impermeable environments: the two-dimensional plane and the three-dimensional space. For classical geometry, there are no "in-betweens." Yet, fractal geometry has forced the breaking of this barrier and poses intermediate dimensions between the 2D plane and the 3D space (section II.6: "Fractals").

This change in vision has not reached clinical medicine. Pathological Anatomy, the clinician's final judge, operates in two perfectly separate modes:

- A three-dimensional "macroscopic mode," operating in the cm range, dealing with perforations, abscesses, atheroma plaques, etc.
- A two-dimensional "microscopic mode," operating in the micron range, referring to atypias, mitosis, lymphocytic infiltrate, etc.

There is no contact bridge between these two approaches. And it is precisely in that "mesoscopic" world where fractal structures operate (and become diseased). The body's fractals are a highly sophisticated architectural way of "unfolding" or "flattening" this dimensional change and allowing the 2D<==>3D transformation.

The absence of "maps" to analyze this fractal world is profoundly dysfunctional, making us blind to some very important physiologic phenomena. There is evidence to suggest that some of the most prevalent pathologies are characterized precisely by the distortion of this architectural framework that allows the dimension change. Some examples include:

Liver Cirrhosis

The liver is a clear example of a fractal structure, in which the portal venous network, the biliary duct network, and the arterial network are exquisitely interwound. The blood from the portal vein passes through a fractal network enveloped by the hepatocyte fractal structure, then converges into the suprahepatic veins that carry it to the heart. This distribution allows an intense metabolic exchange between the blood from the intestine and the hepatocyte cells. Intertwined in this architecture is the fractal structure of the biliary ducts, which converge, draining the bile into the common bile duct and the cystic duct.

One of the typical problems in liver cirrhosis is jaundice, caused by an impediment in the elimination of bilirubin, largely due to structural alteration. But the main cause of morbidity and mortality in liver cirrhosis is portal hypertension.

The distortion of the fractal structure determines an increase in venous pressure in the portal vein (similar to a partially obstructed filter), which is transmitted backwards to the pre-hepatic venous system (producing esophageal varices, with a risk of hemorrhage) and a shunt of blood that bypasses the hepatic filter (conditioning the risk of hepatic encephalopathy). Portal hypertension is probably due to a disorder of the fractal architecture caused by fibrosis and the unstructured regeneration of hepatocytes, but we do not have an instrument to assess (much less quantify) this phenomenon. Although it is a path that is being initiated (Dioguardi et al., 2006; Soda et al., 2003; Wu et al., 1992), it is still outside routine practice and is difficult to perform in percutaneous or transjugular liver biopsy.

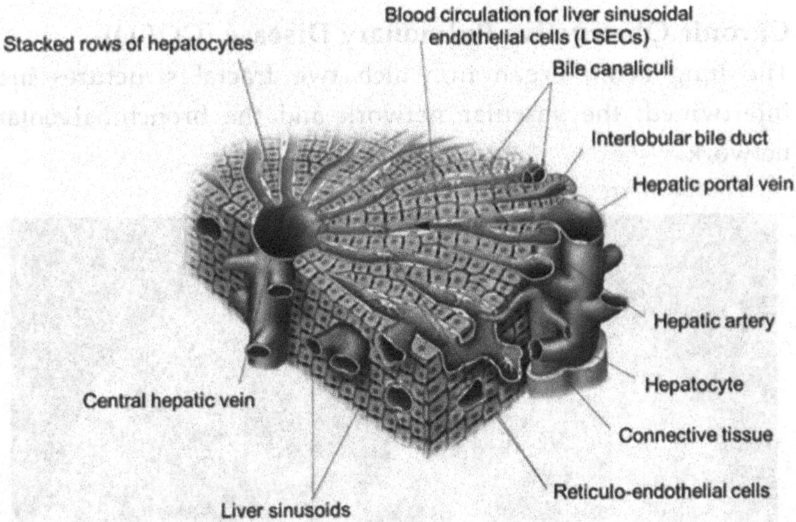

Fig. III.5.4: Microscopic structure of the liver. Image obtained from: Tutty, M.A., Movia, D., Prina-Mello, A. Three-dimensional (3D) liver cell models — a tool for bridging the gap between animal studies and clinical trials when screening liver accumulation and toxicity of nanobiomaterials, *Drug Delivery and Translational Research* (2022) 12:2048–2074, https://doi.org/10.1007/s13346-022-01147-0, under Creative Commons license.

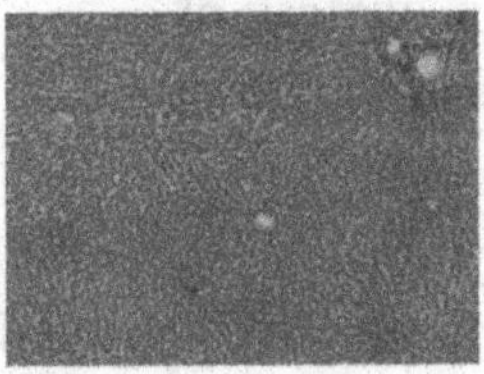

Histology of a normal liver

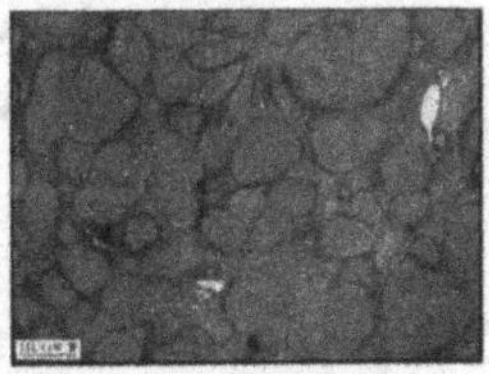

Histology of a cirrhotic liver

Fig. III.5.5: Structural differences between a normal liver and a cirrhotic liver. The presence of fibrotic bands profoundly alters the fractal structure. Images obtained from https://commons.wikimedia.org/wiki/Category:Histology_of_human_liver under Creative Commons license.

Chronic Obstructive Pulmonary Disease (COPD)

The lung is an organ in which two fractal structures are intertwined: the vascular network and the bronchioalveolar network.

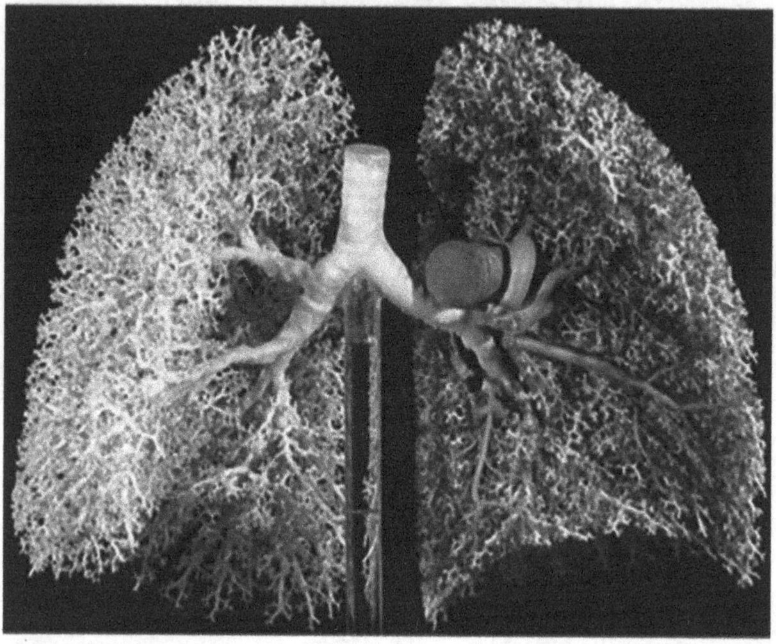

Fig. III.5.6: Fractal structure of the bronchial (right lung) and bronchial + vascular (left lung) trees. Courtesy of Prof. Ewald R. Weibel, from the Institute of Anatomy at the University of Bern.

The distortion of this architecture leads to the two crucial pathophysiological disorders of COPD:

- Pulmonary hypertension, which eventually leads to right heart failure.
- The alteration of the ventilation-perfusion coupling (V/Q mismatch),[1] which produces the hypoxia and hypercapnia associated with COPD.

Once again, conventional pathological anatomy has no tools to measure this architectural distortion. There are indirect ways to evaluate morphologically the degree of pulmonary hypertension (and it can be indirectly assessed by echocardiography or radiology), but the structural alteration cannot be analyzed.

This is even more evident regarding the gasometric alteration. The mismatch is perhaps the central pathophysiological mechanism in COPD. However, since we cannot measure it, it is not considered: it is not taken into account when categorizing COPD patients, and it is hardly discussed in the routine management of patients. There are some attempts to approach an analysis of the fractal structure (Bodduluri et al., 2018; Dunican, 2018; Hwang et al., 2019; Tanabe et al., 2018; Uppaluri et al., 1997; Van de Moortele et al., 2019; Witten et al., 1997), almost always from a radiologic point of view, but Pathological Anatomy, the paradigm of the classic lesional concept, currently has no tools to address this issue in conventional clinical practice.

There are many other diseases which are probably due to a distortion of the fractal architecture (osteoporosis, celiac disease), in which a tool to quantify the architectural alteration would be of great help. But as long as we remain tied to the pathological anatomy map in two "modes" (micro-2D/macro-3D) and cannot explore the mesoscopic fractal deployment, a crucial part of the pathophysiology will escape us.

Footnote

1. The interweaving of the bronchoalveolar fractal structures (coming from the trachea and supplying air) and the vascular structures (arriving through the pulmonary arteries and supplying blood) is exquisitely organized to maximize the exchange of O_2 and CO_2 by evenly distributing air and blood throughout the different regions of the lung. If this architecture is distorted, there will be alveoli that

receive plenty of air (but little blood, losing exchange capacity), and alveoli with plenty of blood and little air (with similar results). Therefore, this balance is crucial to ensure an optimal respiratory function. However, we do not have a tool to measure this coupling. There is a very rough approximation, assuming the existence of a (hypothetical) alveolar space ventilated but not perfused (dead space) and a volume of blood that passes through the lung without contacting the air (shunt effect). In both cases, this reduces the global blood/air interchange and therefore deteriorates the respiratory function.

There are certain techniques to evaluate both the dead space and the shunt effect, but the idea of architectural distortion and the possibility of measuring it objectively is completely lost.

III.6: Sex, Race, Species, Gene

As already discussed, the Map operates in a discrete mode, following an Aristotelian logic of excluded middle ("A is B or not B, there is no other possibility"), whereas the Territory operates in a continuous, analogical, context-dependent manner. Frequently, the transition from one mode to the other involves a collision between these two ways of operating. The present chapter addresses some biological variables commonly considered models of discrete categorization. However, upon closer examination, the edges dissolve and the clear boundary becomes an almost analog fringe.

Sex

At first glance, sex is the archetype of a dichotomous model: male and female. However, upon closer inspection, grays appear. A not inconsiderable number of children are born with ambiguous sexual appearance. It is not clear what volume this population represents (Blackless et al., 2000), but some authors have suggested figures of up to 4% (for comparison, the AB blood group represents 3.5% of the population). Traditionally, this ambiguity was socially unacceptable, and the individual was classified (and eventually surgically intervened) to fit the "Platonic" categories of male or female. It is striking how society accepts significant overlap in secondary sexual characteristics (height, voice, hirsutism, laryngeal prominence), but this is taboo in sexual organs, which determine the assignment to a sex. Perhaps this was due to the traditionally indissoluble relationship between sex and reproductive function.

With the advent of contraceptives, sex became independent of reproduction, with enormous social consequences. Not only does it completely change the population pyramid, but it also allows the emergence of other facets, previously subordinated to the primacy of the reproductive aspect, giving rise to movements like the LGBTQ+ community, which challenges the classic sexual dichotomy.

If we look at other animal species with sexual reproduction, this ambiguity is even more evident. In many animals (many reptiles, some amphibians), sex is not genetically determined but rather dictated by environmental conditions. Some species are strictly hermaphroditic (e.g., *Taenia spp.*). Some animals change sex throughout their lives (for example, the clownfish lives in groups with one male and several females, but when the male disappears, the largest female transforms into a male).

In short, the conventional sexual dichotomy is an example of the "violence" (or at least simplification) that occurs in the transition from Territory to Map, and the consequent "loss of detail" this entails.

Race

This inherent violence in any classification is even more evident when it comes to races. It is obvious that this is a politically charged issue, which has had an evident social control objective. It is also clear that there is a continuum between the different proposed races, and that the dividing line is arbitrary. As a result, in the United States, race is currently defined by the individual themselves (perhaps in part to try to address historical injustices).

However, in Medicine, it is observed that the category "race" (without delving into how it is established) is a factor that significantly correlates with relevant clinical aspects (risks, prognosis, etc.). Some authors suggest that the category "race" captures, albeit imperfectly, genetic differences with possible clinical implications, and that it would be absurd not to use this information. This may be justifiable in descriptive observational studies (although it is well established that the intra-"race" variance is much greater than the inter-"race" variance (AAPA,

1996; Maglo et al., 2016), but it is highly questionable when translated into clinical decisions (Vyas et al., 2020)).

To avoid the enormous problems posed by a formal biological categorization, a more flexible classification has been attempted, focusing mainly on cultural factors. Thus, the concepts of gender (to nuance sex) and ethnicity (to nuance race) emerge, but this destroys the possibility of a strict discrete classification.

Species

The category "species" is the fundamental unit upon which the entire taxonomic classification in Biology is based. Classically, it is defined as the set of organisms or natural populations capable of interbreeding and producing fertile offspring. But once again, the classification system breaks down upon closer examination (Wikipedia, 2020, Species).

- There is a continuum between reproductive isolation and panmixia (complete interbreeding).
- Cryptic species have been described (distinct species, incapable of fertile reproduction, but phenotypically indistinguishable).
- Ring species exist (a set of populations that can interbreed with nearby populations, but not with more distant ones, with which the intermediates can still interbreed).
- In any case, the definition does not apply to asexually reproducing beings.

In this field, the conflict between nominalism and realism, which spans the history of philosophy, is still relevant. "Realist" taxonomists consider that species represent objective leaps in reality (for example, the classification of birds in New Guinea populations in the Stone Age perfectly matches that of current ornithologists), and point out that certain characteristics (e.g., having feathers and laying eggs) are objectively associated. In contrast, "nominalist" taxonomists highlight the impossibility of making categories by formal criteria, and argue that classifications are man-made constructs to help navigate reality. These constructs may be more or less successful (and the only criterion of success is their survival), but they are no more than tentative constructs with a practical purpose. In this dialectic between realism and nominalism, the resonances of the Map/Territory polarity are obvious.

Gene

The gene is the fundamental building block of Genetics. Initially, it was a theoretical construct that allowed for the explanation of some experimental findings. Subsequently, a series of experimental observations solidified it, and the description of chromosomal structure finally established the current conception, formulated around the 1960s. Essentially, a gene is a DNA sequence that encodes a polypeptide molecule. With this, the gene acquires a physical materiality, with a determinable structure and length. The theory of "one gene -> one mRNA -> one protein" was consolidated.

The first problems were solved without major changes to the model: the presence of introns (non-transcribed sequences within a gene), regulatory sequences, or sequences transcribed to RNA but not proteins (mRNA, rRNA, tRNA) were included. But from the 1980s onwards, the model began to shatter (Portin, 2017) (Annex 1).

Naturally, this does not invalidate the concept of the gene. It remains a central instrument as a unit of information (Map): these considerations do not question its utility for mathematical calculations (population genetics) or for medicine (clinical genetics), but the concept of the gene ceases to be univocal and depends heavily on the context. A "political" struggle for the hegemony of its interpretation begins between classical geneticists and developmental biologists (Dietrich, 2000).

In any case, the idea of the gene as the conceptual building block upon which all Biology is based is crumbling: its boundaries become blurry, and its function will depend on a context that is difficult to grasp.

The Dividing Line

In summary, classical classification (Map) becomes blurred as we approach the Territory. However, this boundary is not linear and, as discussed in other chapters, it has characteristics reminiscent of fractal structures. Thus, at some points, it becomes discrete again (with gametes being uniformly dimorphic, one large with nutrients, mitochondria, etc., and the other small and mobile). It becomes "analog" again in the "chromosome Territory" (with crossovers occurring at any point), discrete again in the genetic code territory (involving mutations at specific bases), etc.

In general, it is analog as long as it is structure (Territory), and discrete as long as it deals with information (Map).

Here too, echoes of the yin/yang polarity appear, immiscible but at the same time inseparable, with a Territory/yin mode, analog, context-dependent, and a Map/yang mode, discrete, conceptualizable, and operable.

Annex 1: Problems with the Concept of Gene

Some of the points that problematize the concept of a gene could be:

- Multiple Transcription Start Sites: The same gene can have different starting points for transcription, resulting in different mRNAs and consequently different proteins.
- Alternative Splicing: A gene is transcribed into a pre-mRNA, from which introns (non-coding sequences)

are removed. However, this selective splicing can vary, producing different proteins.

- mRNA Editing: mRNA can be edited post-transcription. Additionally, the same protein can undergo configurational changes affecting its function (protein moonlighting). Therefore, the same gene can produce functionally different proteins.
- There is no precise boundary between transcribed and non-transcribed DNA. Within the cell, there is a continuum of transcribed sequences, making it impossible to define which ones are translated into polypeptide sequences.
- Exon Shuffling: Exons (coding sequences that end up transcribed into amino acids) from different genes can be combined in a single transcript, and there are genes with exons distributed across different chromosomes.
- Genetic Restoration: A phenomenon of genetic restoration occurs more frequently than expected. It is speculated that RNA caches may act as templates to regenerate damaged DNA.
- Epigenetic Inheritance: Most importantly, transgenerational epigenetic inheritance affects gene function. The gene is not the "final word"; it is influenced by its entire context.

III.7: Curves Ahead…

Most of the new technologies in Medicine (genetic, imaging techniques, data science, etc.) derive from the digitization of information; in short, from an "emancipation" of data and the selective extraction of a Map from the Territory. These developments have been driving unprecedented changes in Medicine, and open possibilities that until very recently would have been unimaginable. In any case, they will imply radical transformations in the way medicine is practiced, and probably in the overall conception of the healthcare system. This chapter analyzes some of the possible evolutions.

Crisis of the Ontological Conception of Disease

In III.1: "Medical Nosology" the problems derived from an ontological conception of disease are addressed, and it is proposed how scientific development occurs in a continuous "pathophysiological language" that must be translated (with ever increasing difficulties) to discrete "clinical entities." One of the areas where this is most evident is precisely in the cradle of anatomical-clinical thinking: Pathological Anatomy. A paradigmatic example is lymphomas (tumors of the lymphatic system). What initially was a well-defined clinical picture was undermined in successive classifications (Sternberg-Reed, Rappaport, Kiel, Lukes-Collins, REAL, WHO) initially morphological, then immunophenotypic, and finally also genetic, turning a rare entity into more than 80 different subtypes (Swerdlow et al., 2016), identifiable only by strictly non-clinical criteria.

A similar phenomenon is observed in breast carcinoma. Although the diagnosis is still essentially made in anatomical-clinical language, there are very determining functional factors (hormone receptors, HER2, and Ki-67 expression).

But above all, the prognosis (and therefore the treatment) is increasingly influenced by genetic typing. These techniques analyze the expression of several thousand genes that condition the probabilities of metastasis, recurrence, and response to treatment (Joe, 2022). All this information is increasingly difficult to encapsulate in the classical category system.

Along the same lines, drugs aimed at specific therapeutic targets are radically changing almost all medical specialties (oncology, rheumatology, gastroenterology…). It is not about finding a specific treatment for a disease but about correcting a physiological dysfunction (e.g., achieving terminal differentiation of promyelocytes with tretinoin in acute promyelocytic leukemia (Larson, 2022), or inhibiting the tyrosine kinase receptor with imatinib in certain digestive tumors that present a sustained activation of this receptor (Hemming, 2022)). In short, the consecration of a pathophysiological approach (see III.1: "Medical Nosology") that increasingly misadjusts to the classical anatomical-clinical idea.

Dissolution of the Boundary between Health and Disease

One of the axioms of the pathophysiological conception is that what we call disease is nothing more than a quantitative (not qualitative) change in a variable common to both healthy and sick individuals. This is especially clear in the elderly population, which presents involutional changes (presbycusis, cataracts, osteoporosis, prostatic hyperplasia, cognitive deterioration, etc.), which are difficult to categorize as "disease," but neither are they "healthy."

There are ever more techniques, hitherto reserved for "conventional patients," that could be indicated in a very high percentage of the elderly population (e.g., cochlear implants for the treatment of hypoacusis (Carlson, 2020)), and rejuvenation technologies (e.g., heterochronic parabiosis (Ashapkin et al., 2020)) are on the horizon for which many older people could be candidates.

We are entering a scenario in which the boundary between health and disease is blurred, techniques of *health improvement* appear over a population with an increasingly higher percentage of older patients reaching increasingly older ages (and representing an increasingly larger number of potential voters...). In any case, the disappearance of a clear boundary between health and disease — which establishes who the healthcare system has to take care of — runs the risk of bursting the seams of the model.

And it is curious how this evolution will bring together in the same field the two classic rivals (one on each side) of the "conventional" healthcare model (a Public Health System intended to be universal and based on "Disease Medicine"): on the one hand, the "leftist" version ("hygienist," "Cubanist"), which criticized a model centered on disease and advocated a much more preventive and "social" attitude; on the other hand, the "neoliberal" version, which considered the system inefficient and generating social parasitism, defending the market as the best system for the distribution of commodities (including health). The gods punish men by granting them their wishes: both positions will see their critical predictions fulfilled. What is not clear is that either of them has a solution for the new situation.

Information and Communication Technologies

One of the most evident consequences of technological development is the digitization of data. Information is increasingly independent of a physical medium and is increasingly "in the cloud."

The idea that a patient's medical record is the property of the patient was completely utopian until recently. Now there are multiple platforms where the patient can deposit their data and decide at what time and with whom to share that information. The doctor will be a one-off consultant only to the extent that the patient gives him access.

Medical information (textbooks, journals, clinical trials) is already accessible to anyone interested (and patients are logically very interested in their disease). It is common for physicians to attend to patients who have factual information about their disease equal to or greater than the doctor's and usually closely follow ongoing clinical trials. There are increasingly more applications on the Internet that offer to replace the functions of the doctor. This process also tends towards an increasingly "dislocated" or differently connected medicine. Telemedicine is becoming more frequent, and probably, increasingly specialized surgeries will be performed by robotic and remote means (https://www.intuitive.com/en-us/patients/da-vinci-robotic-surgery). This process of information independence, together with the progressive importance of the patient's right to autonomy, will radically change the function of the conventional healthcare physician. They will have to learn to be a "safe playing-field" where the patient can investigate their situation, helping to contextualize and choose. In short, trying to be more a wise person than a scholar.

Furthermore, transhumanism will imply a radical change of scenario. But this will be analyzed in another chapter.

IV. Echoes and Perspectives

In the previous sections, an attempt has been made to show how the Map/Territory polarity is present in many areas of Western Philosophy and is an urgent problem in several scientific disciplines.

This polarity has also been a central point in some schools of Eastern thought. This section attempts to explore the similarities that may exist between the approach we take in the West and some lines of Eastern thought.

Finally, the transhumanist projects make the Map/Territory polarity the likely most important scientific, philosophical, and moral issue of the coming decades. This section presents some of the positions that have been formulated on this topic.

IV.1: Resonances from the East

In previous chapters, echoes between the Map/Territory polarity and some approaches in Eastern philosophy have been mentioned. This chapter specifically reviews some key points of this polarity and how they are addressed by certain Eastern schools.

Subject/Object Polarity

One of the characteristics of Western science is its (claim to) objectivity. According to what Rafael Echeverría calls the "metaphysical drift" (Echeverría, 2003), it is assumed that there are objects "out there" with their own characteristics (their "essence") which, according to the Newtonian "World-Machine" model, allows for an objective, third-person analysis.

However, this model is breaking down on multiple fronts:

- Quantum physics rules out the possibility of an "external observer." It is inevitable that the observer interferes with the observed. Furthermore, the existence of links between particles (quantum entanglement) incompatible with the "objective" model is demonstrated (see II.4: "Quantum/Classical Duality").
- Nonlinear dynamics ("Chaos Theory") also questions the possibility of a complete objective analysis. The predictions of a model depend on initial conditions that would require infinite precision, which is infeasible in real life.
- Several areas of neuroscience resist a strictly objective approach: perception is necessarily active, and only what is compatible with a prior prediction reaches consciousness (see I.3: "Mind/Map versus Brain/Territory"). On the

other hand, the analysis of consciousness and qualia (the hard problem of neuroscience) seems hardly formulable as "third-person" science (see II.5: "Causality: A Map of the World as a Machine").

- Even from a philosophical point of view, the idea of the "objective" external observer is increasingly disputed (Gabriel, 2015). When someone observes the Universe, from where are they observing it?

This conflict between the observer and the observed is a central part of Zen thought (Suzuki, 1969). For this school, the attempt to describe something objectively, in "third person," automatically fossilizes and destroys the described object, turning it into a corpse.

Taoism holds a similar stance. It understands reality as a global flow (the Tao) where attempting to individualize specific objects or moments is naive (LeGuin, 1997).

Both stances evoke a sort of extreme generalization of the Gaia hypothesis extrapolated to all reality. Any attempt to separate and dissect a part immediately kills and denaturalizes it. Conventional science would assume this is a price to pay for knowledge (once again, the price for moving from Territory to Map), but it is to assume inherently incomplete and fossilized knowledge.

Arguably, the Taoist or Buddhist conceptions are very close to quantum physics (Bohm, 1980; Capra, 1975), and not far from the approach of General Systems Theory (Drack & Wolkenhauer, 2011).

This objectifying impulse of Western science is associated with the conviction that there are "General Laws" of universal compliance, independent of context. And if an exception to one of these laws is observed at some point, it would be due to the existence of an even more General law that would clear the exception (e.g., the relativistic correction to Newtonian physics). At its core, it is the inheritance of a theocratic thought where these laws are given by God (Thomas Aquinas: "a certain Eternal Law, namely, Reason, which exists in the mind of God and governs the whole universe"; Descartes: "laws imposed in nature by God" (quoted by Capra, 1975)). This idea of general laws is so integrated into our scientific culture that it seems obvious.

Eastern culture does not have that origin nor that servitude. It conceives nature as a whole and, rather than obeying Universal Laws, assumes a harmony that only implies the absence of contradiction.

Duality: Conjunction of Opposites

This text has revolved around the idea of two opposite ways of being, schematically referred to as Territory and Map. The attempt to name or define them is already subject to a huge bias because it directly places oneself on one side (Map). Perhaps a more conventional definition of this polarity would be Matter and Information. Throughout the text, multiple examples of this polarity have been presented.

It is striking how this polarity is central to some Eastern schools of thought.

In Taoism, the yin/yang polarity permeates all reality and causes or explains changes (Wang, 2012). Defining yin and yang is not easy and has a significant metaphorical component (probably because, as in the case of Territory/Map, it is done from one of the fields). Originally, it seems that yin/yang referred to the north and south sides of a mountain. Wang cites the most common metaphors as the female/male polarity or darkness/light, or earth/sky. More generally, yin is the contextualizing, rooting, synthetic, heuristic principle: the "Territory" pole *par excellence*. Yang would be the active, focusing, analytical principle corresponding to the "Map" pole.

This polarity is also observed in Hinduism. In this philosophy, *brahman* is the highest universal principle, the Ultimate Reality of the Universe (Lochtefeld, 2002; Otto, 1932), the Territory in its broadest sense. *Atman* would be an individualizing or "essentializing" spirit but actually forms part of a single great soul (*Paramatman*). The relationship between *brahman* and *atman* is one of the central axes of Hindu thought. For conventional science, the path of knowledge would be the elaboration of Maps, ideally increasingly adjusted to the Territory (with decreasing entropy). For the Advaita Vedanta school, enlightenment (*moksha*) is achieved by visualizing the identity of *brahman* and *atman*.

Naturally, the idea of a boundary between Map and Territory is purely metaphorical and cannot be conceived as a spatial boundary. It is important to understand that the Map/Territory conjunction has much more of complementarity than confrontation, and may result from a partial vision. Sometimes a dimensional change can completely modify the landscape: in Fig. IV.1.1, two dark circles are well defined in the 2D plane; however, if the 3D space is considered, they become segments of the same three-dimensional ring.

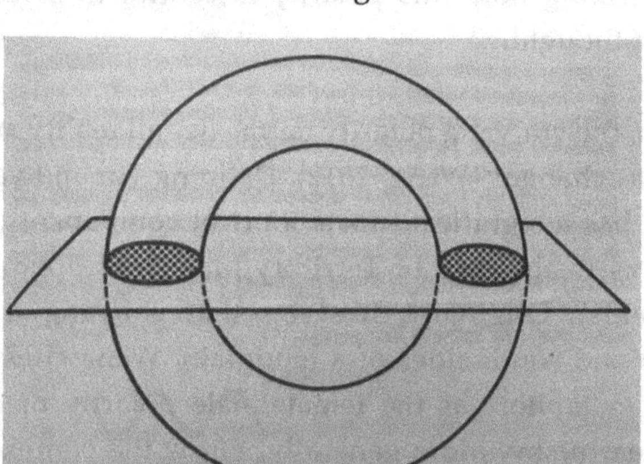

Fig. IV.1.1: "Change of perspective when changing dimension"

There are many seemingly polar binomials that might be interpreted as different aspects of the same reality (mass/ energy, wave/particle, space/time...). This immiscibility and inseparability are analogous to the Taoist yin/yang polarity.

Accepting the spatial metaphor of a boundary, it is striking how it has characteristics reminiscent of a fractal structure: although there is conceptually a clear difference between the polarities, one is always on the cusp, with characteristics common to both poles.

One might argue that a performed musical piece would be Territory (T), while the score would be the map (M). But this boundary is blurry: the original score of Mahler's *2nd Symphony* sold for more than 5 million euros (and in this sense, it is completely "Territory"). In any case, to be performed, the parts for each instrument (M) must be extracted and distributed to the musicians (T).

A computer program (M) needs to exist on a physical substrate (e.g., a computer (T)), which in turn is the concrete realization of a model (M) that someone would draw (T).

Analog sensory stimuli (T) trigger discrete neural responses (firing or not firing) (M). But the neuron's response is not just one-off. The intensity of the stimulus determines the firing frequency (analog) (T), which in turn conditions the response of the next node in the network (M).

It is curious how the Internet, to some extent a kind of meta-Map, ends up becoming Territory: it turns into a Web, over which users navigate.

In summary, although there is a qualitative difference between the two modes (M/T), the border, like in fractal structures, is so interweaved that it is impossible to isolate a pure mode.

States and Movement

The conventional conception considers movement as a process through which certain successive "states" are transited. Although conceptually delicate, this tends to be interpreted as a series of individual "resting mini-states." This is the background of calculus (where the gradient of a curve is established at a point) or kinematics (when the instantaneous speed of a body at time t_1 is calculated). Essentially, it is the interpretation of movement-film as a succession of frames. Once again, it is the leap from the continuous to the discrete, as so often at the Territory/Map frontier.

This idea of a continuum divisible into infinitely small microstates is fraught with problems and has been a central issue for the intuitionist school (Gisin, 2020). It opens the door to paradoxes like Achilles and the tortoise, and leads to a "Platonic world" where all real numbers (with their infinite decimals, never fully explored) are already "created" from the beginning, waiting for us to "discover" them.

Furthermore, the very idea of a "state of rest" is conflictive. There is no "rest," especially when an observer exists. Sound or light involve movement (pressure waves, electromagnetic waves), a temperature above 0°K implies molecular movement, and elementary particles themselves can be interpreted as wave processes.

The rejection of the idea of a state of rest has always been a central point in Taoist thought. This philosophy conceives reality as essentially dynamic. Any attempt at static (punctual) description is inevitably erroneous and distorts an essential part of what it tries to describe.

The Self

Probably, the first leap from Territory to Map is taken by the child between the first and second years of life when they begin to distinguish between self and non-self. This is a crucial difference, indelibly marked on the psyche, and difficult to escape from.

Even in dystopian tales where the possibility is raised that reality might be "fictitious" (e.g., a computer program, as in the movie *The Matrix* (Wachowski & Wachowski, 1999) or the novel *The Anomaly* (Le Tellier, 2020)), it is almost inevitable to anthropomorphize the program (and turn it into a robot, or future descendants implementing a historical research program, etc.). This need to anthropomorphize has significant consequences: it is easy for us to rally against an "evil" (Hitler, Saddam Hussein...), but much harder to confront a "faceless" problem (e.g., climate change).

It is not easy to specify how the "self" is defined. Lakoff (Lakoff & Johnson, 1999) suggests that many of our key ideas are metaphors of physical relationships (I *grasp* your argument, I *see* your point of view, I *understand* (stand under) ... etc.). Possibly the self/non-self dichotomy is a reflection of one of the most primitive metaphors: the inside/outside polarity. That is why it is so difficult for us to accept non-connected entities as objects. It is hard for us to consider the Hanseatic League as a country; we think the relevant objects are the ants (when there are many reasons to think that the significant object is the ant colony (Sasaki & Pratt, 2018)), and the result of distributed intelligence (e.g., the flight of a starling flock) remains completely counterintuitive to us (Flight of the Starlings, n.d.). Arguably the boundary of the self is the body's limits (in Spanish and French, the idea of "putting yourself in someone else's shoes" is translated as "putting yourself in someone else's skin").

In any case, the idea of the self has some clear characteristics:

- It is unquestionable. In our usual state, we think and operate from our self, and it is impossible for us to consider ourselves as nonexistent.

- It is unique and unrepeatable. There is only one self. We cannot imagine that a copy might exist (hence the vertigo in novels like *The Anomaly* (Le Tellier, 2020), or time travel to our past or future).
- It is stable over time. Although we have profoundly changed in every sense, we remain the same self as when we were children and adolescents. Even though the original matter (molecules, cells) has disappeared, we remain the same. This pushes us to consider the self more as an architecture/Map than as an entity/Territory.
- It has the capacity to choose freely. Although we accept a deterministic world (all phenomena have a cause, even if we do not know it), we believe that, with all the limitations, we have the capacity to choose: we are not subject to a strict cause-effect system.

However, this conventional concept of the self is laden with paradoxes. Apart from the need for a "first-person" approach, which is hardly compatible with the conventional scientific method ("hard problem," discussed in I.5: "Consciousness as a Map"), the conventional idea of the self presents many problems.

Its unity/indivisibility is debatable.

- The clinical entity of Multiple Personality Disorder is included in the *Diagnostic and Statistical Manual of Mental Disorders (DSM-5)* (*Diagnostic and Statistical Manual of Mental Disorders*, 5th Edition, 2020), and is defined by the presence of two or more personalities with discontinuity in the sense of self-identity and agency, with changes in affect, behavior, consciousness, memory, perception, cognition, or sensory-motor function.
- The section of the corpus callosum (structure connecting the cortices of the two cerebral hemispheres) produces a

kind of division of the self into two relatively independent identities (Sperry et al., 1969).

- There are intermediate situations (epileptic automatisms, akinetic mutism, asomatognosia, anosognosia), in which it is not clearly definable to what extent there is a conscious self (Damasio, 1999).
- The idea of free will is also questionable. Its more or less general acceptance is striking, in contradiction to the conventional view that all phenomena (except personal decisions?) have an efficient cause. As discussed in II.5: "Causality: A Map of the World as a Machine," there are objective data accessible to an external observer prior to the subject being aware of their decision (Libet, 1985). It seems as if, when we think we make a decision, the decision has already been made.

Criticism of the idea of the self is a very central part of Buddhist philosophy (Gethin, 1998). Buddhism rejects the existence of a self as such. It considers what we call the self to be the fusion of five aggregates (*skandha*) (bodily functions, sensory impressions, conceptualizations, voluntary actions, and consciousness). The conviction that there is a central self, witness and executor of these functions, is, for Buddhists, an illusion. It is striking how this approach parallels the stance of philosopher David Hume (Marías, 1941b; Searle, 2004), who considers that what we call the self is nothing more than the flow of perceptions and contents of consciousness, and it is their uninterrupted continuity that gives the (false) impression of the existence of a self.

This idea of consciousness as a dynamic structure, more organization than object, has significant similarities with natural phenomena such as hurricanes or waves. In both cases, a structure is maintained, but the matter is constantly renewing. The movement of the hurricane or the wave is entirely different from the movement of the air or water molecules that constitute them. They form and dissolve by well-determined physical mechanisms and only exist as long as those mechanisms are maintained (and an observer exists). It makes no sense to ask "where was" the wave or the hurricane before it formed, or "where did it go" upon disappearing.

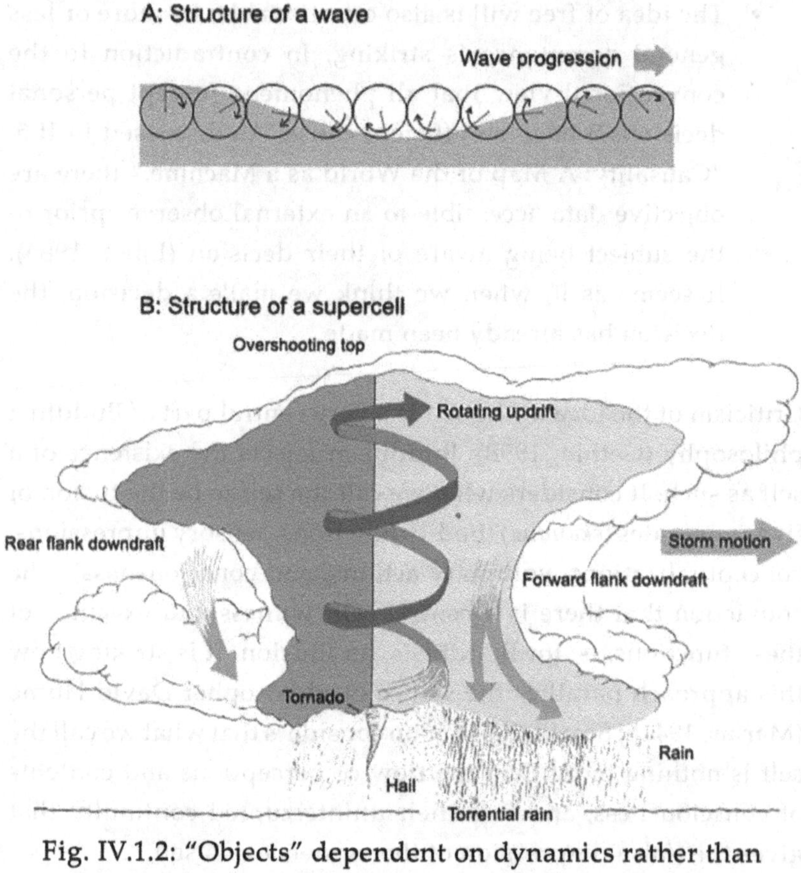

Fig. IV.1.2: "Objects" dependent on dynamics rather than structure. A: Movement of a wave. B: Internal structure of a supercell.

IV.2: Transhumanism, Cis-humanism

As mentioned in I. "Introduction," many authors believe that we are close to being able to emulate consciousness in a computer, or "download" the mind into digital devices. Regardless of the plausibility of the idea or its consequences, it is striking how Medicine has been left out of that debate. More than considering it impossible, it seems as if it hasn't even noticed... This is suicidal. If the expectations of some technicians (Kurzweil, 2010) are met, the physicians currently in their residency training will have to make decisions on the matter, with an absolute lack of specific preparation. In contrast, transhumanism is a hot topic in both the technological and philosophical worlds. Very generally, the positions could be divided into:

- This transition is impossible, at least in the short or medium term. But to argue this position, it is necessary to postulate a "differential trait" of consciousness that is neither structural nor organizational, for which there is little evidence. Generally, these theses end up cornered in principled, almost religious positions.
- This transition will happen in the near future (Tegmark, 2017). It will allow us to expand our potentials and perhaps even offer immortality (Kurzweil, 2010). In any case, our only chance of survival as a species will be to get on that boat (Neuralink, 2018).
- This transition is inevitable, but our nervous system will not be able to modify itself fast enough to keep up (libidinal materialism of Nick Land (*Cybertrop(h)ic*, n.d.)). There will be a "decoupling," and the human race would do well to abandon the "race of reason" and focus on other aspects (emotional, artistic...). This stance aligns

well with fears of a malignant AI, or at least one not aligned with human interests.

An interesting line of inquiry is the work of Joscha Bach and Dietrich Dörner with the PSI program (Bach, 2009). This involves a new architecture of Artificial Intelligence, which aims to model the cognitive process rather than create instruments that emulate brain function. It considers that what we take as characteristically human (emotions, motivation, abstract thinking, symbolic thought, handling uncertainty, generalization, etc.) are algorithms that have been evolutionarily successful. The PSI model attempts to emulate these algorithms, including mechanisms of selection and competition among them. One of the almost inevitable corollaries of this approach is the Bayesian brain model: we do not have access to an "objective reality." We only manipulate scenarios created by us, reaching a balance between computational simplicity and the absence of "dissonances" (mismatch between perceived and predicted). This approach challenges many of our hyperpriors (the idea of objective reality, three-dimensional space, linear time, etc.). One of the points where it diverges from conventional artificial intelligences is its emphasis on the idea of "agency." Conventional AIs learn by establishing increasingly sophisticated relationships (trying to make a Map closer to the Territory: ultimately, trying to "approach the Truth"). Bach suggests that this is a wrong approach: "Objective Truth" is a naive concept, and cognition must be understood always in the context of an action. In this interpretation, memory must be considered only as a part of a "reflex arc" within an action, and the differences between the sensory and motor nervous systems are erased.

An especially interesting alternative is the neorationalism proposed by Negarestani (Negarestani, 2018), which we will discuss below. But to avoid muddying the discussion, some points need to be clarified. The first is the quest for immortality. Nothing against that endeavor, but if included in the discussion, it risks loading it with anxieties and hopes that hinder rational analysis. Another point is the definition of the problem. When we consider duplicating or uploading the mind (the mind, not "me" or "my mind") to a digital system, what do we mean by the mind? This is where Negarestani's proposal becomes relevant.

Neorationalism ("Cis-humanism"?)

Negarestani conceives intelligence as an unstoppable process that tries to make intelligible everything it comes into contact with (including itself). This always "infiltrative" process dismantles the conventional scenario where "intelligence is reached" as a final platform. It is a continuous race of conservation and optimization of its ends, quite similar to the process of infotaxis (active behavior towards obtaining information) that appears in the evolutionary model of cognition proposed by Levin (Scale-Free Cognition (Levin, 2019) also addressed in II.2: "Entropy(ies)": Entropy in Biological Systems). This path, at least as we conceive it, has some clear characteristics:

1.- Discretization

From very early on, to optimize their survival, living beings developed mechanisms to sense and process information from the external environment. This step already involves a process of discretization, transforming the continuous into the discrete. A first step is the differentiation of "self" versus "non-self" (and that is why in biology the development of the membrane is so crucial (see I.3: "Mind/Map versus Brain/Territory": What Do We Mean by "To Exist"?)). But even information processing itself implies discretization. The paradigmatic model of the Turing machine postulates the separation between a moment in which information is "read," and another in which it is processed (and by definition, no other information is received). Therefore, at least in our model of information handling (see Hyperpriors, below) the step from Territory to Map implies a leap from the continuous to the discrete. This process is not a single step. It is a process that repeats itself, one step mounting on top of the other. For example, a first process of discretization would be signs (of alarm, of sexual approach, etc.). These signs carry encapsulated information but are applicable only at that specific point. However, they can be converted into symbols, which are combinable and transferable over time or space. This scale leads to words, concepts, theories, models… Each of these steps, like letters in words, opens an exponentially richer combinatorial world, which would not be computationally possible without the discretization mechanism (see I.1: "Map/ Territory": Advantages of the Map). An evident example is the leap from ideograms (each figure represents a concept) to the alphabet (each figure represents a phoneme, opening a whole world of combinatorics).

2.- *Self-Awareness*

A crucial step in that evolution is when a system becomes aware of itself. This is characteristically a recursive process (being aware of oneself, being aware of being aware of oneself, etc.). This entails the ability to see oneself "from the outside" (i.e., implies the difference between self/non-self, inside/outside) and the ability to imagine how one was before or how one will be later. Therefore, self-awareness is intimately linked to the existence of a History.

All this terrain is treacherous from a terminological point of view. This idea of self-awareness could be similar to that of the Self, mind, and is not far from other terms (spirit, even soul), but they have such an intense ideological load that they become unusable. Perhaps that is why Negarestani chooses the term Geist, in the Hegelian tradition. This Geist, which tries to understand (and understand itself), has clear parallels with our idea of life. In biology, once a replicating system is set in motion (capable of producing copies of itself with sufficient fidelity), it automatically enters the "Darwinian ratchet" (almost perfect replication -> competition with other replicas -> natural selection). This is not a voluntary decision: it is an empirical fact. Likewise, an intelligence that tries to understand (and understand itself) cannot avoid "trying to make its environment intelligible," producing changes that progress that process of understanding. Once again, in the Hegelian tradition, this process of "intelligibilization" of the Geist is nothing other than History.

3.- *Intelligence as Interaction. Computational Trinitarianism*

Although Negarestani does not oppose transhumanism (see below), he is completely skeptical about the idea of "creating" self-awareness. He posits that the process of self-awareness is intimately linked to interaction with others (or other self-awarenesses). Self-awareness ("seeing oneself from the outside") implies accepting that there are other self-awarenesses (that see one, and that one can see). From this perspective, Searle's Chinese room paradox (Cole, 2015) (see I.3: "The Great Journey") dissolves: intelligence is not "in" the room; intelligence is the product of the interaction between the inside and the outside.

Like Christian trinitarianism (three persons, but one true God, accessible from three facets), Negarestani (citing Harper) defends a "computational trinitarianism," with three facets of a single reality: logic, languages, and conceptual structures. The combination of these three aspects (which could also be cited as Mathematics, Syntax, and Semantics) would constitute Computation (or Information).

Negarestani describes how the interaction between intelligences can be analyzed from game theory, and allows the escalation from syntax to semantics to be explained, from the rules of language to the creation of meanings. Therefore, intelligence is inseparable from language ("language is the Dasein of Geist," Negarestani, 2018). Against "expansive" reasoning, which grows "outward" from certain axioms, Negarestani postulates "coherentist" reasoning: there are no axioms, only "truth candidates," which interlock with each other. A candidate will be more "true" the more coherent it is with its context and the better it helps consolidate the conceptual edifice.

Hyperpriors

Just as the idea of life implicitly involves the Darwinian ratchet, the idea of Geist involves the drive to make its environment intelligible, and for that, develop the instruments that best collaborate to that end.

This is one more step in a process that began a long time ago. The first living organisms were hardware capable of reproducing. The development of an increasingly complex nervous system produced systems that, besides reproducing, were capable of modifying their own software (learning, culture). This short-circuited the bottleneck of natural selection based on conventional reproduction, and in a way accelerated evolution from a "Darwinian" pace to a "Lamarckian" pace, being able to modify the phenotype without having to change (and select) the genotype. We are reaching a phase where systems will also be able to modify the hardware. This will likely imply a new evolutionary acceleration, with unpredictable consequences. In this evolution, it is crucial to try to differentiate what are essential characteristics of that process (e.g., constant change) from what may be contingent aspects, linked to the characteristics of the "here and now." For example, it is obvious that humans' image of Reality is crucially conditioned by our special sensitivity to certain frequencies of electromagnetic vibration (visible light: 10^{14} to 10^{15} Hz) and mechanical waves (auditory spectrum: 20 to

20,000 Hz). But there are some "conceptual bricks" that are very difficult for us to overlook ("hyperpriors"). One example would be the concept of space and time: it is extremely difficult for us to reason without the scaffolding of an "objective" space and time, which are, however, increasingly indefensible. Something similar happens with the idea of the Self, or with the need to "discretize" to conceptualize.

Where to Align?

The idea of the Geist as an unstoppable process of making its environment intelligible has profound implications.

It changes the very concept of pedagogy: it is no longer about "teaching" knowledge but about allowing self-development. This trend is evident in the world of computing, with the transition from "imposed" algorithms to self-learning systems (genetic algorithms, artificial intelligence, etc.).

This approach also shapes our attitude toward the possibility of General Artificial Intelligence and represents a new blow against anthropocentric conception. Copernicus and Darwin expelled us from the center of the Universe and the center of Creation. More recently, Dawkins questions the conventional idea that we "have" genes and suggests that, in reality, genes have us: we are the instruments developed by genes to optimize their transmission (Dawkins, 2016). Negarestani makes a similar point. It is naive to suggest that we "possess" intelligence (Geist, spirit, etc.). We are the instruments of the Geist in its development process. And this forces us to reflect on where to align ourselves. Are we the 1.8 kg of brain mass that needs to be protected from A.I., a kind of "geist cancer" that we cannot control? Or do we accept being precisely that developing Geist and understand that the next step may not be the conventional brain, thus assuming that the idea of the classical "self" corresponds to a phase coming to an end? This would mean changing the idea that intelligence serves us to live, to thinking that life is an instrument of intelligence in its development process. The 20th century saw the separation between sex and reproductive function. Will the 21st century see the separation between intelligence and the idea of the Self?

This expansive impulse of the Geist (the Map in the process of becoming one with the Territory) has clear resonances with the Advaita Vedanta school of Hinduism.

Against a transhumanism that considers the essence of the human being to be in self-identity (and probably in a biological substrate), a "cis-humanism" could be claimed, which suggests that what characterizes our humanity is the drive to make intelligible everything we touch. The prefix "trans" implies crossing a barrier (the molecular symmetry axis in chemistry or the sex boundary in gender studies). Transhumanism also implies crossing a barrier (e.g., Kurzweil's singularity, Kurzweil, 2010). Cis-humanism would not see essential changes in a humanism that continued its path as Geist transcending conventional biology and the idea of self-identity. It would not be about opposing "emotions" or "feelings" to Reason but explaining rationally how these phenomena operate and what functions they fulfill (somehow in line with Bach's "Synthetic Intelligence," Bach, 2009). Ultimately, once again, the transition from cause (Territory) to reason (Map).

Quicksand

What are we: our brain or our intelligence? This choice may not be far off and could be posed in concrete and extremely controversial political terrains.

Probably no one seriously questions the political system of a parliamentary democracy. As long as there are no alternatives, it seems extremely irresponsible to challenge the model. But there are profound changes that force reflection.

The idea of democracy is based on two pillars. On the one hand, prudence (a decision made by a majority has less risk of bias or error); on the other hand, justice (all human beings are equal and have the same right to benefit from (and decide on) the measures adopted).

Somehow related to this concept is the idea of the free market, which assumes that each individual is the best to decide where to place his work and money. This translates into the "wise hand" of the market, which would optimize the distribution of goods and services.

These premises are increasingly questionable. It is not proposed to address this issue head-on, but some aspects are unavoidable when considering technological development.

From an economic point of view, the free market showed from the beginning a tendency towards monopolistic concentrations, which then needed to be controlled with corrective mechanisms (Keynesianism). However, progressive technological acceleration has overwhelmed these control mechanisms, and monopolistic concentration (at least in the area of information technologies) has become unstoppable.

Furthermore, it seems increasingly clear that money (the ultimate value of the market system) is also in question. We may find ourselves in a situation like that of colonized African peoples: they were offered trinkets in exchange for a little-valued yellow metal (but crucial for the colonizers). Perhaps money now equates to trinkets, and the new gold is data (which only certain structures can collect and use efficiently).

Technology has also changed the dynamics of the economic system, replacing "human time" with "machine time" that allows for instant transactions, avoiding the theoretical control of the market's "wise hand." In parallel, there has been a "financialization" of the market, making a company's value depend more on its stock market valuation than on its real situation (capital, profits). This shifts priorities and conditions an increasingly unstable market.

The summary of it all is that the "soul" that was at the controls of the market model (individual interest) is increasingly being replaced by algorithmic systems. For now, these systems are supposedly "under human control," but the progressive complexity may eventually end this control.

From a political standpoint, the picture is also not reassuring. The French Revolution led to a parliamentary system that changed the world. Since Napoleon, the "coercive export" of the system (and even more so in colonial regimes) has been conflictive, but currently, it is a model that is difficult to question. Although communism tried to present an alternative (we will not go into how and why it failed), it seems clear that this is a closed experience, and there are no political actors who defend it frontally. In any case, the parliamentary democracy system is currently the only political reference to consider.

However, there is data to suggest it is cracking, and the absence of a clear alternative model does not solve increasingly evident problems.

There is a growing sense that the mechanism of representation is perverted. The system would be parasitized by an endogamic elite that controls it and has managed to turn political life into a show (Debord, 1967), in which participation is a farce.

Moreover, the idea of parliamentary democracy was based on the concept of the Nation-State. Formally, this remains the case, but in reality, the framework of operation has changed. Most major problems cannot be solved within the Nation-State organization, and in any case, there are geopolitical dynamics not subject to democratic control.

Finally, there are global emergencies that the current political system does not seem capable of addressing (climate change, migration crisis, technological revolution, basic income versus full employment debate, etc.).

In short, we face problems of a new level of complexity for which the model inherited from the French Revolution may not be up to the task.

When the time comes, it is easy to imagine an intelligent system offering technically better measures. This is probably already the case in relatively uncontroversial aspects (electricity distribution management, air traffic control). It is not inconceivable that delegating the management of more sensitive issues, such as budget execution, will be proposed. This could occur gradually (they might start with: "you set the 'political' priorities (e.g., balances between freedom/security, equity/autonomy, growth/sustainability, etc.), and the system will optimize its execution"). When this point is reached, we may have to choose between Intelligence (our "natural successor") and our brain (our conventional "Selves").

The democratic system (and, in general, the "New Regime" model) was based on a "Map" created by the Enlightenment and the Encyclopedia. The communist revolution also had its "cartographers" (Marx, Engels, etc.). Perhaps our time is entering a new Territory, which we have not even begun to outline a Map of.

References

AAPA. (1996). AAPA Statement on Race & Racism. *Am J Phys Anthropol*, 101, 569.

Acheson, D. (2017). *The Calculus Story: A Mathematical Adventure*. Oxford University Press.

Adams, R.A., Perrinet, L.U., & Friston, K. (2012). Smooth Pursuit and Visual Occlusion: Active Inference and Oculomotor Control in Schizophrenia. *PLoS ONE*. https://doi.org/10.1371/journal.pone.0047502

Adams, R.A., Stephan, K.E., Brown, H.R., Frith, C.D., & Friston, K.J. (2013). The Computational Anatomy of Psychosis. *Frontiers in Psychiatry*. https://doi.org/10.3389/fpsyt.2013.00047

Al-Khalili, J., & McFadden, J. (2014). *Life on the Edge: The Coming of Age of Quantum Biology*. Crown Publishers.

Arendt, L.M., Rudnick, J.A., Keller, P.J., & Kuperwasser, C. (2010). Stroma in breast development and disease. *Seminars in Cell and Developmental Biology*. https://doi.org/10.1016/j.semcdb.2009.10.003

Ashapkin, V., Kutueva, L.I., & Vanyushin, B.F. (2020). The Effects of Parabiosis on Aging and Age-Related diseases. *Advances in Experimental Medicine and Biology*. https://doi.org/10.1007/978-3-030-42667-5_5

Ashford Lee, E. (2017). Evolution and Revolution, Models in Crisis. In *Plato and the Nerd: The Creative Partnership of Humans and Technology* (pp. 115–122). MIT Press.

Bach, J. (2009). *Principles of Synthetic Intelligence: PSI: An Architecture of Motivated Cognition*. Oxford University Press.

Baer, Y. (1978). *A History of the Jews in Christian Spain*. The Jewish Publication Society of America, 1, 301.

Ballow, M., & Shehata, N. (2022). Overview of intravenous immune globulin (IVIG) therapy. *UpToDate*.

Benjamin, W. (1936). *El narrador*. Metales Pesados.

Bergson, H. (1927). *Essai sur les données immédiates de la conscience* (10th edition). Presse Universitaires de France.

Berkeley, G. (1734). The Analyst: a Discourse Addressed to an Infidel Mathematician. https://www.maths.tcd.ie/pub/HistMath/People/Berkeley/Analyst/Analyst.pdf

Berlin, I., Pensamiento, T., & Marí, S. (1999). *Las raices del romanticismo*. Pensamiento/Taurus Series. Taurus.

Blackless, M., Charuvastra, A., Derryck, A., Fausto-Sterling, A., Lauzanne, K., & Lee, E. (2000). How sexually dimorphic are we? Review and synthesis. *Am J Hum Biol*, 12(2), 151–166. https://doi.org/10.1002/(SICI)1520-6300(200003/04)12:2<151::AID-AJHB1>3.0.CO;2-F

Bodduluri, S., Puliyakote, A.S.K., Gerard, S.E., Reinhardt, J.M., Hoffman, E.A., Newell Jr., J.D., Nath, H.P., Han, M.K., Washko, G.R., San José Estépar, R., Dransfield, M.T., Bhatt, S.P., & COPD Gene Investigators. (2018). Airway fractal dimension predicts respiratory morbidity and mortality in COPD. *J Clin Invest*, 128(12), 5374–5382. https://doi.org/10.1172/JCI120693

Bohm, D. (1980). *Wholeness and the Implicate Order*. Routledge.

Bowles, S. (2011). Cultivation of cereals by the first farmers was not more productive than foraging. *Proceedings of the National Academy of Sciences of the United States of America*. https://doi.org/10.1073/pnas.1010733108

Brewer, J.A., Worhunsky, P.D., Gray, J.R., Tang, Y.Y., Weber, J., & Kober, H. (2011). Meditation experience is associated with differences in default mode network activity and connectivity. *Proceedings of the National Academy of Sciences of the United States of America*. https://doi.org/10.1073/pnas.1112029108

Bronowski, J. (1978). *El sentido común de la ciencia*. Ediciones Península.

Büchel, C., Geuter, S., Sprenger, C., & Eippert, F. (2014). Placebo analgesia: A predictive coding perspective. *Neuron*. https://doi.org/10.1016/j.neuron.2014.02.042

Buckner, R.L., & Carroll, D.C. (2007). Self-projection and the brain. *Trends in Cognitive Sciences*. https://doi.org/10.1016/j.tics.2006.11.004

Buonomano, D. (2018). *Your Brain Is a Time Machine: The Neuroscience and Physics of Time*. W.W. Norton & Company.

Burdette, S., & Schwartz, R.S. (1987). Idiotypes and Idiotypic Networks. *New England Journal of Medicine*. https://doi.org/10.1056/nejm198707233170407

Burr, D.C., Morrone, M.C., & Ross, J. (1994). Selective suppression of the magnocellular visual pathway during saccadic eye movements. *Nature*. https://doi.org/10.1038/371511a0

Capra, F. (1975). *The Tao of Physics: An Exploration of the Parallels between Modern Physics and Eastern Mysticisms* (5th edition). Shambhala Publications, Inc.

Carhart-Harris, R.L. (2018). The entropic brain – revisited. *Neuropharmacology*. https://doi.org/10.1016/j.neuropharm.2018.03.010

Carhart-Harris, R.L., & Friston, K.J. (2010). The default-mode, ego-functions and free-energy: a neurobiological account of Freudian ideas. *Brain*, 133(4), 1265–1283. https://doi.org/10.1093/brain/awq010

Carhart-Harris, R.L., & Friston, K.J. (2019). REBUS and the anarchic brain: Toward a unified model of the brain action of psychedelics. *Pharmacological Reviews*. https://doi.org/10.1124/pr.118.017160

Carhart-Harris, R.L., Leech, R., Hellyer, P.J., Shanahan, M., Feilding, A., Tagliazucchi, E., Chialvo, D.R., & Nutt, D. (2014). The entropic brain: A theory of conscious states informed by neuroimaging research with psychedelic drugs. *Frontiers in Human Neuroscience*, 8(1 FEB), 1–22. https://doi.org/10.3389/fnhum.2014.00020

Carlson, M.L. (2020). Cochlear Implantation in Adults. *New England Journal of Medicine*. https://doi.org/10.1056/nejmra1904407

Carrasco, J. (1989). *El método estadístico en la investigación médica* (4th edition). Editorial Ciencia 3 SA.

Carrére, E. (2021). *Yoga*. Anagrama.

Carru, C., Pes, G.M., Deiana, L., Baggio, G., Franceschi, C., Lio, D., Balistreri, C.R., Candore, G., Colonna-Romano, G., & Caruso, C. (2003). Association between the HFE mutations and longevity: a study in Sardinian population. *Mech Ageing Dev,* 124(4), 529–532. https://doi.org/10.1016/s0047-6374(03)00032-0

Chaitin, G. (2005). *Meta Math!: The Quest for Omega*. Pantheon Books.

Churruca, J., Vigil, L., Luna, E., Ruiz-Galiana, J., & Varela, M. (2008). The route to diabetes: Loss of complexity in the glycemic profile from health through the metabolic syndrome to type 2 diabetes. *Diabetes Metab Syndr Obes,* 1, 3–11. http://www.ncbi.nlm.nih.gov/pubmed/21437151

Clark, A. (2016a). Gaze Allocation: Doing What Comes Naturally. In *Surfing Uncertainty* (pp. 66–68).

Clark, A. (2016b). "Surfing Uncertainty." *TLS – The Times Literary Supplement*. https://doi.org/10.1093/acprof:oso/9780190217013.001.0001

Cole, D. (2015). The Chinese Room Argument (Edward N. Zalta & Uri Nodelman, eds.). *The Stanford Encyclopedia of Philosophy*; Metaphysics Research Lab, Stanford University. https://plato.stanford.edu/archives/win2015/entries/chinese-room/

Cooper Jr., L.T. (2009). Medical Progress: Myocarditis. *New England Journal of Medicine.*

Cybertrop(h)ic. (n.d.). The Engineer & the Artist: Negarestani versus Land. *Cybertrop(h)ic*. https://cybertrophic.wordpress.com/2021/02/13/the-engineer-the-artist-negarastani-vs-land/

Damasio, A. (1999). *The Feeling of What Happens: Body, emotion and the making of consciousness*. Harcourt Brace & Company.

Damasio, A. (2010). *Self Comes to Mind: Constructing the Conscious Brain*.

Damasio, A.R. (2006). *Descartes' Error*. Random House.

Dauben, J.W. (1977). Georg Cantor and Pope Leo XIII: Mathematics, Theology, and the Infinite. *Journal of the History of Ideas*. https://doi.org/10.2307/2708842

Dawkins, R. (2016). *The Selfish Gene*. Oxford University Press.

de Caso, I. (2022). *Psicodélicos y salud mental: La neurociencia de una nueva psicoterapia*. Ciencia Psicodélica.

De Ridder, D., Vanneste, S., & Freeman, W. (2014). The Bayesian brain: Phantom percepts resolve sensory uncertainty. *Neuroscience and Biobehavioral Reviews*. https://doi.org/10.1016/j.neubiorev.2012.04.001

Debord, G. (1967). *La société du spectacle*. Essaie Folio.

DeDeo, S. (n.d.). Introduction to Renormalization. Santa Fe Institute. Retrieved September 4, 2021, from https://www.complexityexplorer.org/courses/67-introduction-to-renormalization

Diagnostic and Statistical Manual of Mental Disorders, 5th Edition. (2020). In Definitions. https://doi.org/10.32388/taiky6

Diamond, F. (2017). *Guns, Germs and Steel: A short history of everybody for the last 13,000 years*. W.W. Norton & Co.

Dietrich, M.R. (2000). The problem of the gene. *C R Acad Sci III*, 232(12), 1139–1146.

Dioguardi, N., Grizzi, F., Franceschini, B., Bossi, P.,. & Russo, C. (2006). Liver fibrosis and tissue architectural change measurement using fractal-rectified metrics and Hurst's exponent. *World J Gastroenterol*, 12(14), 2187–2194. https://doi.org/10.3748/wjg.v12.i14.2187

Distante, S., Robson, K.J., Graham-Campbell, J., Arnaiz-Villena, A., Brissot, P., & Worwood, M. (2004). The origin and spread of the *HFE*-C282Y haemochromatosis mutation. *Hum Genet*, 115(4), 269–279. https://doi.org/10.1007/s00439-004-1152-4

Drack, M., & Wolkenhauer, O. (2011). System approaches of Weiss and Bertalanffy and their relevance for systems biology today. *Semin Cancer Biol*, 21(3), 150–155. https://doi.org/10.1016/j.semcancer.2011.05.001

du Sautoy, M. (2007). *La música de los números primos: El enigma de un problema matemático abierto*. Acantilado.

Dunican, E.M. (2018). Seeing the forest for the trees: fractal dimensions measure COPD airway remodeling. *J Clin Invest*, 128(12), 5203–5205. https://doi.org/10.1172/JCI124776

Dyck, E. (2005). Flashback: Psychiatric experimentation with LSD in historical perspective. *Canadian Journal of Psychiatry*. https://doi.org/10.1177/070674370505000703

Echeverría, R. (2003). *Ontología del Lenguaje*. Comunicaciones Noreste Ltda.

Edwards, M.J., Adams, R.A., Brown, H., Pareés, I., & Friston, K.J. (2012). A Bayesian account of "hysteria." *Brain*. https://doi.org/10.1093/brain/aws129

Einstein, A., & Infeld, L. (1938). *The Evolution of Physics: From Early Concepts to Relativity and Quanta*. Simon & Schuster.

Faber, K. (1930). *Nosography: The Evolution of Clinical Medicine in Modern Medicine*. Paul B. Hoeber.

Fair, D.A., Cohen, A.L., Dosenbach, N.U.F., Church, J.A., Miezin, F.M., Barch, D.M., Raichle, M.E., Petersen, S.E., & Schlaggar, B.L. (2008). The maturing architecture of the brain's default network. *Proceedings of the National Academy of Sciences of the United States of America*. https://doi.org/10.1073/pnas.0800376105

Fisher, S., & Marcus, G. (2006). The eloquent ape: genes, brains and the evolution of language. *Nat Rev Genet* 7, 9–20.

Fitzgerald, D.W., Sterling, T.R., & Haas, D.W. (2020). Mycobacterium tuberculosis. In J.E. Bennet, R. Dolin, & M.J. Blaser (eds.), Mandell, Douglas, and Bennett's *Principles and Practice of Infectious Diseases* (9th edition). Elsevier.

Fletcher, P.C., & Frith, C.D. (2009). Perceiving is believing: a Bayesian approach to explaining the positive symptoms of schizophrenia. *Nature Reviews Neuroscience*. https://doi.org/10.1038/nrn2536

Flight of the Starlings. (n.d.). *National Geographic.* Retrieved February 28, 2022, from https://www.youtube.com/watch?v=V4f_1_r80RY

Fox, M.D., Snyder, A.Z., Vincent, J.L., Corbetta, M., Van Essen, D.C., & Raichle, M.E. (2005). The human brain is intrinsically organized into dynamic, anticorrelated functional networks. *Proceedings of the National Academy of Sciences of the United States of America.* https://doi.org/10.1073/pnas.0504136102

Friston, K. (2010). The free-energy principle: a unified brain theory? *Nature Reviews Neuroscience.* https://doi.org/10.1038/nrn2787

Gabriel, M. (2015). *Por qué el mundo no existe.* Pasado y Presente.

Gao, W., Zhu, H., Giovanello, K.S., Smith, J.K., Shen, D., Gilmore, J.H., & Lin, W. (2009). Evidence on the emergence of the brain's default network from 2-week-old to 2-year-old healthy pediatric subjects. *Proceedings of the National Academy of Sciences of the United States of America.* https://doi.org/10.1073/pnas.0811221106

Genesis. 2:5–6, New International Version.

Georgiadis, J.R., & Kringelbach, M.L. (2012). The human sexual response cycle: Brain imaging evidence linking sex to other pleasures. *Progress in Neurobiology.* https://doi.org/10.1016/j.pneurobio.2012.05.004

Gethin, R. (1998). No Self: Personal Continuity and Dependent Arising. In *The Foundations of Buddhism* (pp. 133–162). Oxford University Press.

Gisin, N. (2020). Mathematical languages shape our understanding of time. *Nature Physics,* 16, 114–116. https://doi.org/10.1038/s41567-019-0748-5

Goldberger, A.L. (1996). Non-linear dynamics for clinicians: chaos theory, fractals, and complexity at the bedside. *Lancet,* 347(9011), 1312–1314. https://doi.org/10.1016/s0140-6736(96)90948-4

Goldberger, A.L., Rigney, D.R., & West, B.J. (1990). Chaos and fractals in human physiology. *Sci Am*, 262(2), 42–49. https://doi.org/10.1038/scientificamerican0290-42

Goody, J. (1977). *The Domestication of the Savage Mind*. Cambridge University Press.

Greene, B. (2004). *The Fabric of the Cosmos: Space, Time and the Texture of Reality*. Vintage Books.

Grob, C.S., Danforth, A.L., Chopra, G.S., Hagerty, M., McKay, C.R., Halberstad, A.L., & Greer, G.R. (2011). Pilot study of psilocybin treatment for anxiety in patients with advanced-stage cancer. *Archives of General Psychiatry*. https://doi.org/10.1001/archgenpsychiatry.2010.116

Hall, J.E., and Hall, M.E. (2021). *Guyton and Hall Textbook of Medical Physiology*. Elsevier.

Hanahan, D., & Weinberg, R.A. (2000). The Hallmarks of Cancer. *Cell*. https://doi.org/10.1016/S0092-8674(00)81683-9

Hanzlíková, Z., Kofler, M., Slovák, M., Věchetová, G., Fečíková, A., Kemlink, D., Sieger, T., Růžička, E., Valls-Solé, J., Edwards, M.J., & Serranová, T. (2019). Prepulse inhibition of the blink reflex is abnormal in functional movement disorders. *Movement Disorders*. https://doi.org/10.1002/mds.27706

Harari, Y.N. (2016). *Homo Deus: A Brief History of Tomorrow*. Random House.

Harley, J.B. (2001). *The New Nature of Maps: Essays in the History of Cartography*. Johns Hopkins University Press.

Harris, D.J., Vine, S.J., & Wilson, M.R. (2017). Neurocognitive mechanisms of the flow state. *Progress in Brain Research*. https://doi.org/10.1016/bs.pbr.2017.06.012

Helbert, M. (2017). Antibody Diversity. In *Immunology for Medical Students* (pp. 35–43). Elsevier.

Hemming, J. (2022). Tyrosine kinase inhibitor therapy for advanced gastrointestinal stromal tumors. *UpToDate*.

Henrich, J. (2020). *The Weirdest People in the World: How the West Became Psychologically Peculiar and Particularly Prosperous.* Farrar, Straus and Giroux.

Hirsch, I.B. (2022). Pathogenesis of type 1 diabetes mellitus. *UpToDate.*

Hoffman, D. (2015). *Do we see reality as it is?* (TED, ed.). https://www.youtube.com/watch?v=oYp5XuGYqqY

Hoffman, D.D. (2016). The Evolutionary Argument Against Reality. *Quanta Magazine.*

Huang, Y.L., & Jones, B. (1982). On the interdependence of temporal and spatial judgments. *Perception & Psychophysics.* https://doi.org/10.3758/BF03204862

Hwang, J., Oh, Y.M., Lee, M., Choi, S., Seo, J.B., Lee, S.M., & Kim, N. (2019). Low morphometric complexity of emphysematous lesions predicts survival in chronic obstructive pulmonary disease patients. *Eur Radiol,* 29(1), 176–185. https://doi.org/10.1007/s00330-018-5551-7

Ishihara, M., Keller, P.E., Rossetti, Y., & Prinz, W. (2008). Horizontal spatial representations of time: Evidence for the STEARC effect. *Cortex.* https://doi.org/10.1016/j.cortex.2007.08.010

Jackson, F. (1986). What Mary Didn't Know. *Journal of Philosophy,* 83, 291–295.

Jeffery, K., Pollack, R., & Rovelli, C. (2019). On the Statistical Mechanics of Life: Schrödinger Revisited. *Entropy,* 21(12), 1211. https://doi.org/10.3390/e21121211

Joe, B. (2022). Clinical features, diagnosis, and staging of newly diagnosed breast cancer. H. Burstein & S. Vora (eds.), *UpToDate.*

Kahneman, D. (2012). *Pensar rápido, pensar despacio.* Debate.

Kandel, E. (2012). The Emergence of Twentieth-Century Painting. In *The Age of Insight: The Quest to Understand the Unconscious in Art, Mind, and Brain.* Random House USA.

Kasemeier-Kulesa, J.C., Teddy, J.M., Postovit, L., Seftor, E.A., Seftor, R.E.B., Hendrix, M.J.C., & Kulesa, P.M. (2008). Reprogramming multipotent tumor cells with the embryonic neural crest microenvironment. *Developmental Dynamics*, 237(10), 2657–2666. https://doi.org/10.1002/dvdy.21613

Knights, A.J., Funnell, A.P.W., Crossley, M., & Pearson, R.C.M. (2012). Holding Tight: Cell Junctions and Cancer Spread. *Trends in Cancer Research*, 8, 61–69.

Krebs, T.S. & Johansen, P.Ør. et al. (2012). Lysergic acid diethylamide (LSD) for alcoholism: meta-analysis of randomized controlled trials. *Journal of Psychopharmacology*. https://doi.org/10.1177/0269881112439253

Kuhn, T.S. (1996). *The Structure of Scientific Revolutions* (3rd edition). University of Chicago Press.

Kuriyama, S. (1999). *The Expressiveness of the Body and the Divergence of Greek and Chinese Medicine*. Zone Books.

Kurzweil, R. (2010). *The Singularity is Near*. Gerald Duckworth & Co.

Lai, M., & Chopra, S. (2020). Hepatitis A virus infection in adults: Epidemiology, clinical manifestations, and diagnosis. *UpToDate*.

Lain Entralgo, P. (1982). *El diagnóstico médico: Historia y teoría*. Salvat Editores.

Lakoff, G., & Johnson, M. (1999). *Philosophy in the Flesh: The Embodied Mind and Its Challenge to Western Thought*. Basic Books.

Lakoff, G., & Johnson, M. (2013). *Metaphors We Live By*. University of Chicago Press. https://doi.org/10.7208/chicago/9780226470993.001.0001

Larson, R. (2022). Initial treatment of acute promyelocytic leukemia in adults. *UpToDate*.

Le Tellier, H. (2020). *L'anomalie*. Gallimard.

LeGuin, U.K. (1997). *Lao Tzu: Tao Te Ching. A Book about the Way and the Power of the Way*. Shambhala Publications, Inc.

Leibniz, G. (1857). *Nouvelles Lettres et Opuscules inédits de Leibniz* (F. de Careil, ed.; p. 234).

Lenfant, C. (2003). Shattuck lecture–clinical research to clinical practice–lost in translation? *New England Journal of Medicine*, 349(9), 868–874.

Levin, M. (2019). The Computational Boundary of a "Self": Developmental Bioelectricity Drives Multicellularity and Scale-Free Cognition. *Frontiers in Psychology*, 10. https://doi.org/10.3389/fpsyg.2019.02688

Levin, M. (2021). Bioelectric signaling: Reprogrammable circuits underlying embryogenesis, regeneration, and cancer. *Cell*, 184(8), 1971–1989. https://doi.org/10.1016/j.cell.2021.02.034

Levin, M. (2023). Darwin's agential materials: evolutionary implications of multiscale competency in developmental biology. *Cellular and Molecular Life Sciences*, 80(6), 142. https://doi.org/10.1007/s00018-023-04790-z

Lévy, T. (1987). *Figures de l'infini. Les mathématiques au miroir des cultures*. Editions du Seuil.

Libet, B. (1985). Unconscious cerebral initiative and the role of conscious will in voluntary action. *Behavioral and Brain Sciences*. https://doi.org/10.1017/S0140525X00044903

Lio, D., Balistreri, C.R., Colonna-Romano, G., Motta, M., Franceschi, C., Malaguarnera, M., Candore, G., & Caruso, C. (2002). Association between the MHC class I gene HFE polymorphisms and longevity: a study in Sicilian population. *Genes & Immunity*, 3(1), 20–24. https://doi.org/10.1038/sj.gene.6363823

Lipsitz, L.A., & Goldberger, A.L. (1992). Loss of "complexity" and aging. Potential applications of fractals and chaos theory to senescence. *JAMA*, 267(13), 1806–1809. http://www.ncbi.nlm.nih.gov/pubmed/1482430

Lochtefeld, J.G. (2002). *The Illustrated Encyclopedia of Hinduism* (Issue 1). The Rosen Publishing Group.

Lok, Anna S.F. (2022). Characteristics of the hepatitis B virus and pathogenesis of infection. *UpToDate*.

Lundelin, K., Vigil, L., Bua, S., Gomez-Mestre, I., Honrubia, T., & Varela, M. (2010). Differences in complexity of glycemic profile in survivors and nonsurvivors in an intensive care unit: A pilot study. *Critical Care Medicine*, 38(3). https://doi.org/10.1097/CCM.0b013e3181ce49cf

Maglo, K.N., Mersha, T.B., & Martin, L.J. (2016). Population genomics and the statistical values of race: an interdisciplinary perspective on the biological classification of human populations and implications for clinical genetic epidemiological research. *Front Genet*, 7, 22.

Mandelbrot, B. (1967). How long is the coast of Britain? Statistical self-similarity and fractional dimension. *Science*, 156(3775), 636–638. https://doi.org/10.1126/science.156.3775.636

Mandelbrot, B. (1983). *The Fractal Geometry of Nature* (W.H.F. and Company, ed.). Henry Holt and Company.

Mann, S.F., Pain, R., & Kirchhoff, M.D. (2022). Free energy: a user's guide. *Biology & Philosophy*, 37(4), 33. https://doi.org/10.1007/s10539-022-09864-z

Mann, T. (1924). *La montaña mágica*. Edhasa, 2015.

Marías, J. (1971a). El Idealismo alemán. In *Historia de la Filosofía* (23rd edition, pp. 275–328). Revista de Occidente.

Marías, J. (1971b). El empirismo. La filosofía inglesa. Hume. In *Historia de la Filosofía* (23rd edition, p. 249). Revista de Occidente.

Milner, J.C. (2013). *Por una política de los seres hablantes: Breve tratado político* 2. Grama Ediciones.

Moreno, F.A., Wiegand, C.B., Taitano, E.K., & Delgado, P.L. (2006). Safety, tolerability, and efficacy of psilocybin in 9 patients with obsessive-compulsive disorder. *Journal of Clinical Psychiatry*. https://doi.org/10.4088/JCP.v67n1110

Neel, J.V. (1999). Diabetes mellitus: a "thrifty" genotype rendered detrimental by "progress"? 1962. *Bull World Health Organ*, 77(8), 693–694.

Negarestani, R. (2018). *Intelligence and Spirit*. Urbanomic/ Sequence Press.

Neuralink. (2018). *Neuralink* (Volume 2019). https://www.neuralink.com/

Nietzsche, F. (1872). *El nacimiento de la tragedia: o Grecia y el pesimismo*. Alianza Editorial.

Nisbett, R.E., Choi, I., Peng, K., & Norenzayan, A. (2001). Culture and systems of thought: Holistic versus analytic cognition. *Psychological Review*. https://doi.org/10.1037/0033-295X.108.2.291

Nishimura, T., Tokuda, I., Miyachi, S. et al. (2022). Evolutionary loss of complexity in human vocal anatomy as an adaptation for speech. *Science*, 377(6607), 760–763.

Otto, R. (1932). *Mysticism East and West: A Comparative Analysis of the Nature of Mysticism*. Macmillan and Co.

Pagel, M. (2017). Q&A: What is human language, when did it evolve and why should we care? *BMC Biology*. https://doi.org/10.1186/s12915-017-0405-3

Pellicano, E., & Burr, D. (2012). When the world becomes "too real": A Bayesian explanation of autistic perception. *Trends in Cognitive Sciences*. https://doi.org/10.1016/j.tics.2012.08.009

Popper, K., & Eccles, J. (1977). *El yo y su cerebro*. Labor Universitaria.

Portin, P., & Wilkins, A. (2017). The Evolving Definition of the Term "Gene." *Genetics*, 205(4), 1353–1364.

Prindle, A., Liu, J., Asally, M., Ly, S., Garcia-Ojalvo, J., & Süel, G.M. (2015). Ion channels enable electrical communication in bacterial communities. *Nature*, 527(7576), 59–63. https://doi.org/10.1038/nature15709

ProfessionalAI.com. (2020). Deep Blue Algorithm. https://www.professional-ai.com/deep-blue-algorithm.html

Raichle, M.E., & Snyder, A.Z. (2007). A default mode of brain function: A brief history of an evolving idea. *NeuroImage*. https://doi.org/10.1016/j.neuroimage.2007.02.041

Rees, T., Bosch, T., & Douglas, A.E. (2018). How the microbiome challenges our concept of self. *PLoS Biology*. https://doi.org/10.1371/journal.pbio.2005358

Richardson, L.F. (1961). The Problem of Contiguity: An appendix to statistics of deadly quarrels. In *General Systems: Yearbook of the Society for the Advancement of General Systems Theory* (Vol. 6, pp. 139–187).

Roberts, W.C. (2001). High salt intake, its origins, its economic impact, and its effect on blood pressure. *Am J Cardiol*, 88(11), 1338–1346. https://doi.org/10.1016/s0002-9149(01)02105-1

Roche, D., & Mahon, P. (2021). Depth of Anesthesia Monitoring. *Anesthesiology Clinics*. https://doi.org/10.1016/j.anclin.2021.04.004

Rodríguez de Castro, C., Vigil, L., Vargas, B., García Delgado, E., García Carretero, R., Ruiz-Galiana, J., & Varela, M. (2017). Glucose time series complexity as a predictor of type 2 diabetes. *Diabetes/Metabolism Research and Reviews*, 33(2). https://doi.org/10.1002/dmrr.2831

Rosch, E. (1978). Principles of categorization. In E. Rosch & B. Lloyd (eds.), *Cognition and Categorization*. Lawrence Erlbaum Associates.

Rossouw, J.E., Anderson, G.L., Prentice, R.L., LaCroix, A.Z., Kooperberg, C., Stefanick, M.L., Jackson, R.D., Beresford, S.A.A., Howard, B.V., Johnson, K.C., Kotchen, J.M., & Ockene, J. (2002). Risks and Benefits of Estrogen Plus Progestin in Healthy Postmenopausal Women: Principal Results from the Women's Health Initiative Randomized Controlled Trial. *Journal of the American Medical Association*. https://doi.org/10.1001/jama.288.3.321

Rovelli, C. (2019). *The Order of Time*. Penguin Books.

Sackett, D.L., Rosenberg, W.M.C., Gray, J.A.M., Haynes, R.B., & Richardson, W.S. (1996). *Evidence-Based Medicine: How to practice and teach EBM*. Churchill Livingstone.

Sasaki, T., & Pratt, S.C. (2018). The Psychology of Superorganisms: Collective Decision Making by Insect Societies. *Annual Review of Entomology*. https://doi.org/10.1146/annurev-ento-020117-043249

Schrödinger, E. (1944). *What is Life?* Cambridge University Press.

Searle, J. (2004). *Mind: A Brief Introduction*. Oxford University Press.

Seth, A. (2021). *Being You: A New Science of Consciousness*. Faber & Faber Ltd.

Silverstein, A.M. (1989). *A History of Immunology*. Academic Press Inc.

Sobel, D. (2006). *Longitud*. Anagrama.

Soda, G., Nardoni, S., Bosco, D., Grizzi, F., Dioguardi, N., & Melis, M. (2003). Fractal analysis of liver fibrosis. *Pathologica*, 95(2), 98–102. http://www.ncbi.nlm.nih.gov/pubmed/12768879

Sperry, R.W., Gazzaniga, M.S., & Bogen, J.E. (1969). Interhemispheric relationships: the neocortical commissures; syndromes of hemisphere disconnection. In P.J. Vinken and G.W. Bruyn (eds.), *Handbook of Clinical Neurology*.

Steinberg, K.K., Cogswell, M.E., Chang, J.C., Caudill, S.P., McQuillan, G.M., Bowman, B.A., Grummer-Strawn, L.M., Sampson, E.J., Khoury, M.J., & Gallagher, M.L. (2001). Prevalence of C282Y and H63D mutations in the hemochromatosis (HFE) gene in the United States. *JAMA*, 285(17), 2216–2222. https://doi.org/10.1001/jama.285.17.2216

Suzuki, D.T. (1969). *The Field of Zen* (C. Humphreys, ed.). Harper & Row.

Swerdlow, S.H., Campo, E., Pileri, S.A., Lee Harris, N., Stein, H., Siebert, R., Advani, R., Ghielmini, M., Salles, G.A., Zelenetz,

A.D., & Jaffe, E.S. (2016). The 2016 revision of the World Health Organization classification of lymphoid neoplasms. *Blood*. https://doi.org/10.1182/blood-2016-01-643569

Tanabe, N., Muro, S., Sato, S., Oguma, T., Sato, A., & Hirai, T. (2018). Fractal analysis of low attenuation clusters on computed tomography in chronic obstructive pulmonary disease. *BMC Pulm Med*, 18(1), 144. https://doi.org/10.1186/s12890-018-0714-5

Tegmark, M. (2017). *Life 3.0: Being human in the Age of Artificial Intelligence*. Knopf.

Tononi, G., & Cirelli, C. (2006). Sleep function and synaptic homeostasis. *Sleep Medicine Reviews*. https://doi.org/10.1016/j.smrv.2005.05.002

Uppaluri, R., Mitsa, T., Sonka, M., Hoffman, E.A., & McLennan, G. (1997). Quantification of pulmonary emphysema from lung computed tomography images. *Am J Respir Crit Care Med*, 156(1), 248–254. https://doi.org/10.1164/ajrccm.156.1.9606093

Urban, T. (2017). *Neuralink and the Brain's Magical Future*. https://waitbutwhy.com/2017/04/neuralink.html

Van de Moortele, T., Goerke, U., Wendt, C.H., & Coletti, F. (2019). Airway morphology and inspiratory flow features in the early stages of Chronic Obstructive Pulmonary Disease. *Clin Biomech* (Bristol, Avon), 66, 60–65. https://doi.org/10.1016/j.clinbiomech.2017.11.005

van den Beld, A., & Lamberts, S. (2021). Endocrinología y envejecimiento. In S. Melmed et al. (eds.), *Williams. Tratado de Endocrinología* (14th edition, pp. 1179–1194). Elsevier.

van der Linden, D., Tops, M., & Bakker, A.B. (2021). Go with the flow: A neuroscientific view on being fully engaged. *European Journal of Neuroscience*. https://doi.org/10.1111/ejn.15014

Van Essen, D.C., & Dierker, D.L. (2007). Surface-based and probabilistic atlases of primate cerebral cortex. *Neuron*. https://doi.org/10.1016/j.neuron.2007.10.015

Varela, M. (1991). *La crisis del concepto de entidad morbosa. Consideraciones sobre nuestra nosotaxia*. Facultad de Medicina: Vol. Tesis Doct. Universidad Autónoma de Madrid.

Varela, M., Churruca, J., Gonzalez, A., Martin, A., Ode, J., & Galdos, P. (2006). Temperature curve complexity predicts survival in critically ill patients. *American Journal of Respiratory and Critical Care Medicine,* 174(3). https://doi.org/10.1164/rccm.200601-058OC

Varela, M., Rodriguez, C., Vigil, L., Cirugeda, E., Colas, A., & Vargas, B. (2015). Glucose series complexity at the threshold of diabetes. *Journal of Diabetes,* 7(2). https://doi.org/10.1111/1753-0407.12182

Varela, M., Ruiz-Esteban, R., & Mestre De Juan, M.J. (2010). Chaos, fractals, and our concept of disease. *Perspectives in Biology and Medicine,* 53(4).

Varela, M., Vigil, L., Rodriguez, C., Vargas, B., & Garciá-Carretero, R. (2016). Delay in the Detrended Fluctuation Analysis Crossover Point as a Risk Factor for Type 2 Diabetes Mellitus. *Journal of Diabetes Research*. https://doi.org/10.1155/2016/9361958

Vargas, A.S., Luís, Â., Barroso, M., Gallardo, E., & Pereira, L. (2020). Psilocybin as a new approach to treat depression and anxiety in the context of life-threatening diseases — a systematic review and meta-analysis of clinical trials. *Biomedicines*. https://doi.org/10.3390/biomedicines8090331

Vargas, B., & Varela, M. (2013). Facing a Shift in Paradigm at the Bedside? *International Journal of Clinical Medicine,* 04(01), 35–40. https://doi.org/10.4236/ijcm.2013.41008

Vargas, B., Cuesta-Frau, D., Ruiz-Esteban, R., Cirugeda, E., & Varela, M. (2015). What can biosignal entropy tell us about health and disease? Applications in some clinical fields. *Nonlinear Dynamics, Psychology, and Life Sciences,* 19(4).

Vyas, D.A., Eisentsein, L.G., & Jones, D.S. (2020). Hidden in Plain Sight — Reconsidering the Use of Race Correction in

Clinical Algorithms. *New England Journal of Medicine*, 383, 874–882. https://doi.org/10.1056/NEJMms2004740

Wachowski, L., & Wachowski, L. (1999). *The Matrix*.

Wang, R.R. (2012). *Yinyang: The Way of Heaven and Earth in Chinese Thought and Culture*. Cambridge University Press.

Weiden, P.L., Sullivan, K.M., Flournoy, N., Storb, R., Thomas, E.D., & the Seattle Marrow Transplant Team. (1981). Antileukemic effect of chronic graft-versus-host disease: contribution to improved survival after allogeneic marrow transplantation. *New England Journal of Medicine*, 304(25), 1529–1533. https://doi.org/10.1056/NEJM198106183042507

Weinberg, H.L. (1991). *Levels of Knowing and Existence: Studies in General Semantics*. Harper.

Whiten, A. (2021). The burgeoning reach of animal culture. *Science*. https://doi.org/10.1126/SCIENCE.ABE6514

Wikipedia. Chaos theory. https://en.wikipedia.org/wiki/Chaos_theory

Wikipedia. Species. https://en.wikipedia.org/wiki/Species

Wikipedia. Pierre-Simon Laplace. https://en.wikipedia.org/w/index.php?title=Pierre-Simon_Laplace&oldid=800201933

Witten, M.L., Tinajero, J.P., Sobonya, R.E., Lantz, R.C., Quan, S.F., & Lemen, R.J. (1997). Human alveolar fractal dimension in normal and chronic obstructive pulmonary disease subjects. *Res Commun Mol Pathol Pharmacol*, 98(2), 221–230. http://www.ncbi.nlm.nih.gov/pubmed/9467830

Wu, C.M., Chen, Y.C., & Hsieh, K.S. (1992). Texture features for classification of ultrasonic liver images. *IEEE Trans Med Imaging*, 11(2), 141–152. https://doi.org/10.1109/42.141636

Yamasaki, H. et al. (1995). Intercellular communication and carcinogenesis. *Mutation Research/Fundamental and Molecular Mechanisms of Mutagenesis*, 333(1–2), 181–188. https://doi.org/10.1016/0027-5107(95)00144-1

Zalta, E., Nodelman, U., Allen, C., Kim, H., & Oppenheimer, P. (2021). Infinity. *Stanford Encyclopedia of Philosophy*. https://plato.stanford.edu/entries/infinity/

Zhou, M., Zheng, M., Zhou, X., Tian, S., Yang, X., Ning, Y., Li, Y., & Zhang, S. (2023). The roles of connexins and gap junctions in the progression of cancer. *Cell Communication and Signaling*, 21(1), 8. https://doi.org/10.1186/s12964-022-01009-9

Index

IFF BOOKS

ACADEMIC AND SPECIALIST

Framespotting

Changing how you look at things changes how you see them

Laurence & Alison Matthews

A punchy, upbeat guide to framespotting. Spot deceptions and hidden assumptions; swap growth for growing up. See and be free.

Paperback: 978-1-78279-689-3 ebook: 978-1-78279-822-4

Is There an Afterlife?

David Fontana

Is there an Afterlife? If so what is it like? How do Western ideas of the afterlife compare with Eastern? David Fontana presents the historical and contemporary evidence for survival of physical death.

Paperback: 978-1-90381-690-5

Nothing Matters

a book about nothing

Ronald Green

Thinking about Nothing opens the world to everything by illuminating new angles to old problems and stimulating new ways of thinking.

Paperback: 978-1-84694-707-0 ebook: 978-1-78099-016-3

Panpsychism

The Philosophy of the Sensuous Cosmos

Peter Ells

Are free will and mind chimeras? This book, anti-materialistic but respecting science, answers: No! Mind is foundational to all existence.

Paperback: 978-1-84694-505-2 ebook: 978-1-78099-018-7

Punk Science
Inside the Mind of God
Manjir Samanta-Laughton
Many have experienced unexplainable phenomena; God, psychic abilities, extraordinary healing and angelic encounters. Can cutting-edge science actually explain phenomena previously thought of as 'paranormal'?
Paperback: 978-1-90504-793-2

The Vagabond Spirit of Poetry
Edward Clarke
Spend time with the wisest poets of the modern age and of the past, and let Edward Clarke remind you of the importance of poetry in our industrialized world.
Paperback: 978-1-78279-370-0 ebook: 978-1-78279-369-4
